Proof and the Art of Mathematics

Proof and the Art of Mathematics

Joel David Hamkins

The MIT Press
Cambridge, Massachusetts
London, England

This book was set using LaTeX and TikZ by the author. Printed and bound in the United States of America.

Library of Congress Cataloging-in-Publication Data is available.

ISBN: 978-0-262-53979-1

10 9 8 7 6 5 4 3

To my students—may all their theorems be true, proved by elegant arguments that flow effortlessly from hypothesis to conclusion, while revealing fantastical mathematical beauty.

Contents

Preface

This is a mathematical coming-of-age book, for students on the cusp, who are maturing into mathematicians, aspiring to communicate mathematical truths to other mathematicians in the currency of mathematics, which is: proof. This is a book for students who are learning—perhaps for the first time in a serious way—how to write a mathematical proof. I hope to show how a mathematician makes an argument establishing a mathematical truth.

Proofs tell us not only that a mathematical statement is true, but also why it is true, and they communicate this truth. The best proofs give us insight into the nature of mathematical reality. They lead us to those sublime yet elusive *Aha!* moments, a joyous experience for any mathematician, occurring when a previously opaque, confounding issue becomes transparent and our mathematical gaze suddenly penetrates completely through it, grasping it all in one take. So let us learn together how to write proofs well, producing clear and correct mathematical arguments that logically establish their conclusions, with whatever insight and elegance we can muster. We shall do so in the context of the diverse mathematical topics that I have gathered together here in this book for the purpose.

What is a proof, really? Mathematicians are sometimes excused from jury duty, it is said, because according to the prosecutors, they do not know what it means to prove something "beyond a reasonable doubt," which is the standard of evidence that juries follow for conviction in US criminal courts. Indeed, a mathematician's standard of evidence for proof is very high, perhaps too high for the prosecutors to want them on the jury.

Mathematical logicians have a concept of formal proof, which is a detailed form of proof written in a rigid formal language. These proofs, often intended to be processed computationally, understood and verified by a machine, can irrefutably establish the validity of the theorems they prove, yet they are often essentially unreadable by humans, usually providing us with little mathematical insight beyond the truth of the raw statement that is proved. Despite this, the emerging field of automated theorem proving may in future decades profoundly transform mathematical practice.

In high-school geometry, students often learn a standard two-column form of proof, in which certain kinds of statements are allowed in the left column, provided that they are justified by admissible reasons in the right column. This form of proof highlights an all-important feature, namely, that proofs must provide a chain of reasoning logically establishing their conclusion from their premises.

Meanwhile, in contemporary mathematical research writing, one finds a more open, flexible proof format, written essay style in complete sentences while retaining the defining feature that a proof must logically establish the truth of its conclusion. In fact, this style of proof writing has endured millennia. If you open Euclid's *Elements*, for example, now translated into nearly every human language, you will find beautiful proofs written in an engaging open-prose style of argument. The same proof-writing style permeates mathematical writing through the ages, up to the present day, filling our mathematical research journals and books.

It is this style of proof writing with which we shall be concerned in this book, the open-prose style of argument that mathematicians use to establish and communicate to one another the truth of a mathematical claim. For the purposes of this book, let me say:

> A *proof* is any sufficiently detailed convincing mathematical argument that logically establishes the conclusion of a theorem from its premises.

Mathematical proofs naturally make use of a variety of proof methods, which may vary depending on the mathematical topic. We shall undertake proofs in number theory, for example, using the principle of mathematical induction; in combinatorics, using various counting arguments; and in game theory, graph theory, real analysis, and so on, using methods common to these diverse areas. Throughout, we shall learn about certain common pitfalls for beginners and emphasize new aspects of proof writing when they arise. There are, of course, a variety of proof styles that one may adopt within the theorem-proof format that I describe and use in this book, and I hope that the reader may grow to develop his or her own mathematical voice.

If I had to describe my own proof-writing style, I would say that I am somewhat less concerned with providing all of the technical details of an argument, especially when these can be easily provided by the reader or when they obscure or distract from the idea of the proof, and somewhat more concerned with providing a deeper explanation, with conveying the essential idea of an argument in such a way as to cultivate mathematical insight and understanding in my reader. I often strive to lead the reader through a series of smaller or plainly stated mathematical truths that in combination make the larger conclusion clear. To my way of thinking, a key value of many mathematical achievements consists in their making certain formerly difficult ideas easy to understand. Ultimately, my aim is mathematical insight; I aspire to induce an *Aha!* moment in my reader. For this reason, I prefer to write in plain language, when I can, or to explain a mathematical idea in ordinary words, when this successfully conveys the intended mathematical idea. But of course, sometimes the only way to convey a mathematical idea accurately is with detailed mathematical notation or formalism, and in this case one must step up with the right tools. So although I strive for nontechnical simplicity, I do not use this as an excuse for vagueness, and naturally I strive to provide fuller details or technical notation when the clarity of the argument requires it.

This book was typeset using LaTeX. Except for the two images on pages 41 and 146, which are in the public domain, and the image of the Königsberg bridges on page 133, which I drew myself by hand, I created all the other images in this book using TikZ in LaTeX, specifically for this book.

Joel David Hamkins
July 2019

A Note to the Instructor

In this book, I have assembled a collection of what I find to be compelling mathematical statements with interesting elementary proofs, illustrating diverse proof methods and intended to develop a beginner's proof-writing skills. All who aspire toward mathematics, who want to engage fully with the mathematical craft by undertaking a mathematical analysis and constructing their own proofs of mathematical statements, will benefit from this text, whether they read it as part of a university proof-writing course or study it on their own.

I should like to emphasize, however, that the book is not an axiomatic development of its topics from first principles. The reason is that, while axiomatic developments certainly involve proof writing, I find that they are also often burdened, especially in their beginnings, with various tedious matters. Think of the need, for example, to establish the associativity of integer addition from its definition. I find it sensible, in contrast, to separate the proof-writing craft in its initial or introductory stages from the idea that an entire mathematical subject can be developed from weak axiomatic principles. I also find it important to teach proof writing with mathematically compelling, enjoyable examples, which can inspire a deeper interest in and curiosity about mathematics; students will then be motivated to work through other examples on their own.

So the proofs in this book are not built upon any explicitly given list of axioms but, rather, appeal to very general mathematical principles with which I expect the reader is likely familiar. My hope is that students, armed with the proof-writing skills they have gained from this text, will go on to undertake axiomatic treatments of mathematical subjects, such as number theory, algebra, set theory, topology, and analysis.

The book is organized around mathematical themes, rather than around methods of proof, such as proofs by contradiction, proofs by cases, proofs of if-then statements, or proofs of biconditionals. To my way of thinking, mathematical ideas are best conceived of and organized mathematically; other organizational plans would ultimately be found artificial. I do not find proofs by contradiction, for example, to be a natural or robust mathematical category. Such a proof, after all, might contain essentially the same mathematical insights and ideas as a nearby proof that does not proceed by contradiction. Indeed, mathematical statements will often admit diverse proofs, using diverse proof methods—I explore this point explicitly in chapter 2—and it would seem wrong to me to suggest that a given statement should be proved only with a particular method.

For these reasons, I have organized the book around a selection of mathematically rich topics, having interesting theorems with elementary proofs. I have chosen what I find to be especially enjoyable mathematical topics—at times fanciful, yet always amenable to a serious mathematical analysis—because I feel students will learn best how to write proofs with material that is itself intrinsically interesting. Please enjoy! I have simply treated the various proof methods as they arise naturally within each topic. Some of the topics, such as the theory of games and order theory, seem to be somewhat less often covered in the contemporary mathematics curriculum than I believe is warranted; therefore, kindly let the students learn some of this beautiful mathematics while developing their proof-writing skills.

If the book is to be used as part of an undergraduate proof-writing course, then each chapter could be the basis of a lecture or sometimes several lectures. Some chapters are a little longer, and the earlier chapters are generally a little easier. The chapters on Pick's theorem (chapter 8) and on the polygonal dissection congruence theorem (chapter 10) are each essentially chapter-length developments of a single major theorem. The final chapters, on infinity (chapter 13), order theory (chapter 14), and real analysis (chapter 15), are perhaps a little more abstract or difficult.

Very little of the material depends fundamentally on earlier chapters, except that the method of mathematical induction, covered in chapter 4, is used in many arguments throughout the book, and some of the material in chapters 14 and 15 depends on chapter 13 and on the properties of functions in chapter 11. I would find it sensible to cover relations (chapter 11) before graph theory (chapter 12).

I would recommend that instructors cover any or all of the topics they find appealing, starting with chapter 1. The students can be directed to read the preface and other preliminary material on their own, especially "A Note to the Student," but the theorem-proof format should be discussed explicitly in class, probably in the first lecture. Students can also be instructed to peruse the mathematical habits from later chapters, as they are applicable generally. When I used this book as the textbook for an undergraduate proof-writing course at the City University of New York, we covered the entire book in one fifteen-week semester.

I recommend that essentially all homework and exam work for a proof-writing course using this textbook require the student to write proofs. The students should prove essentially every substantive mathematical statement they make. I have included exercises in each chapter, and I would find it normal for students to attempt every exercise in any chapter that is covered. A few sample answers are provided at the end of the book, and a more extensive volume of answers is available from MIT Press. Students should be encouraged to form study groups, to work on the exercises together, since one learns mathematics very effectively by trying out one's ideas on one's peers. In this sense, the practice of mathematics can be considered a social performance activity. Please encourage your students to understand the mathematical idea in the proofs they study, rather than to memorize specific proofs. Indeed, I find little value in having students memorize specific mathematical proofs; in my experience anyone who understands a proof well can reliably reproduce it, even months or years later, without ever having specifically memorized it.

I also recommend that the instructor spend time in class giving careful solutions to some or even all of the exercises, with fully detailed proofs, after the students have attempted them, so that the students can benefit from exposure to further examples of sound mathematical proof. Please do not worry about "giving away the idea," since surely there is an infinite supply of new mathematical questions with which the students might struggle on their own. In my experience and perception of our subject, what students seem to need most is exposure to far more and better examples of high-quality mathematical arguments. Give them an abundant supply of well-written, well-explained proofs in your lectures and in your office hours; I think you will find that they pick it up by example.

A Note to the Student

Welcome to mathematical proof writing. I hope you enjoy the mathematical ideas I have assembled for you in this book; I tried to include topics I find especially engaging. I want to help you learn how to communicate mathematics as a mathematician does, which is to say, with mathematical proof. Let us learn together how to write proofs well.

I encourage all beginning proof writers to adopt the *theorem-proof* format of proof writing in all their mathematical work, including all the proof writing they may undertake with this book. This format will help you organize and clarify your thoughts and stay on track with a correct mathematical argument. To use the format, one begins by writing the word "Theorem" (or Lemma, Corollary, and so on, as appropriate) and then writing a clear and precise statement of the mathematical claim that is being made.

> **Theorem.** *A clear and precise statement of the mathematical claim.*
>
> *Proof.* A logically correct, clear, and precise argument that establishes the truth of the claim made in the theorem statement. □

The formulation of a precise statement to be proved is a critical step in proof writing, and the theorem-proof format requires one to state one's claims explicitly. Beginners are sometimes sloppy or neglectful about this, embarking on a "proof" without being sufficiently clear about exactly what it is they are proving. But such a practice can lead to error; truth is slowly buried in the shifting sands of what is being claimed and what is being proved. Make it clear what you are proving by formulating a precise theorem statement.

After formulating a clear theorem statement, the theorem-proof format calls for you to write the word "Proof," followed by a logical argument establishing the truth of the theorem, marking the end of the proof with □, a demarcation that helps organize and clarify your discourse as to whether you are giving an example, motivating an analysis, or giving a proof.

In time, one expands the repertoire of the types of formal statements to include definitions, lemmas, and corollaries. The general flow is something like this:

Definition. A clear and precise statement giving the official meaning of a mathematical term.

Lemma. *A small theorem-like result, which will be proved separately and then used in other proofs.*

Proof. Each lemma requires its own proof.   □

Theorem. *A clear and precise statement.*

Proof. A logically correct, clear, and precise argument establishing the theorem statement.   □

Corollary. *A theorem-like result that follows easily from a previously proved theorem or from details arising in a previous proof.*

Proof. Each corollary should be proved, even if the proof is merely a short argument pointing out that the corollary follows from a previous theorem.

□

Each definition is a formal statement describing once and for all the meaning that will be ascribed to a given mathematical term. In any mathematical argument, this term shall have exactly the meaning described in the definition—nothing more, nothing less. Beginning mathematicians may find themselves consulting and reconsulting the definitions of the mathematical terms appearing in their arguments, to make sure that they are not making unjustified assumptions. My practice is to give major concepts their own formally set-out and numbered definitions, while other minor terms are defined within the flow of the ordinary text.

A theorem is a formal statement of a mathematical claim that will be proved; theorems are often considered as the principal unit of mathematical advance, encapsulating a concise mathematical truth. The equally important proof of the theorem consists of a clear and concise mathematical argument that logically justifies the truth of the statement made in the theorem. A lemma is something like a small theorem, a convenient fact or major step that is good to know already during the course of a much longer proof. A lemma might be reused several times. For some theorems, it is convenient during the proof to first prove a lemma or a sequence of lemmas and only afterward embark on the main argument. A corollary is a theorem-like statement that can be easily proved as a consequence of a previous theorem.

Please use the theorem-proof format of writing even when solving the exercises in this text. For example, suppose the exercise says:

Exercise. Prove that every hibdab is hobnob.

Then you should begin your solution not by starting directly with your proof, or by rewriting those instructions. Rather, you begin by writing a clear statement of what you are proving, like this:

Theorem. *Every hibdab is hobnob.*

Proof. And so on with your argument.   □

Notice that this turns the instruction statement of the exercise into a new, clear mathematical statement. It would have made no sense to prove the original assertion, "Prove that every hibdab is a hobnob," because that is not a mathematical statement, a statement that might be true or false, but is rather an imperative, an *instruction* about what we should do. We carry out that instruction by formulating a clear mathematical statement as our theorem and then proving this statement. In this way, you shall turn every exercise into your own formal theorem statement and proof.

Let me say lastly that I have also gathered together in this text a collection of what I find to be sound mathematical habits of mind, bits of mathematical wisdom or advice that I believe to be beneficial or even fundamental to sound mathematical practice, highlighted in boxes at the end of each chapter. Adopting these habits, I believe, will help an aspiring mathematician solve a problem, find an elusive proof, or write better proofs. Let me mention one of them right now that we have just discussed.

Use the theorem-proof format. In all your mathematical exercises, write in the theorem-proof style. State a clear claim in your theorem statement. State lemmas, corollaries, and definitions as appropriate. Give a separate, clearly demarcated proof for every formally stated mathematical claim.

About the Author

I am an active research mathematician working on the mathematics and philosophy of the infinite. I have published about one hundred research articles on diverse topics in mathematical logic and set theory, ranging from forcing and large cardinals to infinitary computability and infinite game theory, including infinite chess. My blog, *Mathematics and Philosophy of the Infinite*, features a variety of mathematical posts and commentary, including an accessible series on "Math for Kids," and the reader can discover what pricks my mathematical fancy on Twitter (see links below). In addition, I have posted over one thousand research-level mathematical arguments on MathOverflow, the popular mathematics question-and-answer forum, which is becoming a fundamental tool for mathematical research. Interested readers can therefore find many of my mathematical proofs and arguments online. For me, mathematics has been an enjoyable lifelong learning process; I continually strive to improve.

I am also a mathematical philosopher, working in mathematical and philosophical logic with a focus on infinity, especially in set theory, the philosophy of set theory, and the philosophy of mathematics. In recent work, I have been exploring pluralism in the foundations of mathematics by introducing and investigating the multiverse view in set theory and the mathematics and philosophy of potentialism. My forthcoming book *Lectures on the Philosophy of Mathematics*, upon which I have based my lectures here in Oxford, emphasizes a mathematically grounded perspective on the subject.

I have taught college-level mathematics for over twenty-five years, mostly at the City University of New York, where I have taught essentially every proof-based undergraduate topic in pure mathematics, as well as many graduate-level topics, especially in logic. At CUNY I was Professor of Mathematics, Professor of Philosophy, and Professor of Computer Science.

I have recently and quite happily taken up a new position as Professor of Logic at the University of Oxford, where I am a member of the Faculty of Philosophy, affiliated member of the Mathematics Institute, and the Sir Peter Strawson Fellow in Philosophy at University College. I have also held diverse visiting professor positions over the years, at New York

University, Carnegie Mellon University, the University of Cambridge, the University of Toronto, Kobe University, the University of Amsterdam, the University of Münster, the University of California at Berkeley, and elsewhere. I earned my B.S. in mathematics (1988) from the California Institute of Technology and my Ph.D. in mathematics (1994) at the University of California at Berkeley.

Joel David Hamkins
Professor of Logic and Sir Peter Strawson Fellow in Philosophy
Oxford University, University College
High Street, Oxford OX1 4BH

joeldavid.hamkins@philosophy.ox.ac.uk
joeldavid.hamkins@maths.ox.ac.uk
Blog: http://jdh.hamkins.org
MathOverflow: http://mathoverflow.net/users/1946/joel-david-hamkins
Twitter: @JDHamkins

1 A Classical Beginning

One of the classical gems of mathematics—and to my way of thinking, a pinnacle of human achievement—is the ancient discovery of incommensurable numbers, quantities that cannot be expressed as the ratio of integers.

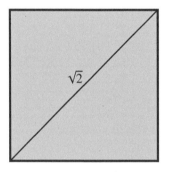

The Pythagoreans discovered in the fifth century BC that the side and diagonal of a square have no common unit of measure; there is no smaller unit length of which they are both integral multiples; the quantities are *incommensurable*. If you divide the side of a square into ten units, then the diagonal will be a little more than fourteen of them. If you divide the side into one hundred units, then the diagonal will be a little more than 141; if one thousand, then a little more than 1414. It will never come out exactly. One sees those approximation numbers as the initial digits of the decimal expansion:

$$\sqrt{2} = 1.4142135623730950488016887242096980785\ldots$$

The discovery shocked the Pythagoreans. It was downright heretical, in light of their quasi-religious number-mysticism beliefs, which aimed to comprehend all through proportion and ratio, taking numbers as a foundational substance. According to legend, the man who made the discovery was drowned at sea, perhaps punished by the gods for impiously divulging the irrational.

1.1   The number $\sqrt{2}$ is irrational

In modern terminology, the claim is that $\sqrt{2}$ is irrational, since this is the ratio of diagonal to side in any square, and the *rational* numbers are those expressible as a fraction of integers p/q, with $q \neq 0$. So we may state the claim as follows.

Theorem 1. *The number $\sqrt{2}$ is irrational.*

In plainer words, the theorem asserts that $\sqrt{2}$ cannot be expressed as a fraction. Let us prove the theorem.

Proof of theorem 1. Suppose toward contradiction that $\sqrt{2}$ is rational. This means that $\sqrt{2} = p/q$ for some integers p and q, with $q \neq 0$. We may assume that p/q is in lowest terms, which means that p and q have no nontrivial common factors. By squaring both sides, we conclude that $2 = p^2/q^2$ and consequently $2q^2 = p^2$. This tells us that p^2 must be an even number, which implies that p also is even, since the square of an odd number is odd. So $p = 2k$ for some integer k. From this, it follows that $2q^2 = p^2 = (2k)^2 = 4k^2$, and therefore $q^2 = 2k^2$. So q^2 also is even, and therefore q must be even. Thus, both p and q are even, which contradicts our assumption that p/q was in lowest terms. So $\sqrt{2}$ cannot be rational.   □

That was a traditional argument, and my impression is that if you were to knock on doors down the hall of a university mathematics department asking for a proof that $\sqrt{2}$ is irrational, then first of all, every single mathematician you find will be able to prove it and second, the odds are good for you to get a version of that argument. To my way of thinking, just as every self-respecting educated person should know something of the works of Shakespeare and of the heliocentric theory of planetary motion, similarly everyone should be able to prove the irrationality of the square root of 2. Mathematicians now have dozens of proofs of this beautiful classical result—the reader can find various collections online—and by the end of this chapter, we shall see several.

Notice that the proof proceeded *by contradiction*. In such a proof, one assumes ("toward contradiction") that the desired conclusion is not true, aiming to derive from this assumption a contradiction, an impossible consequence. Since that consequence cannot be the true state of affairs, the assumption leading to it must have been wrong, and so the desired conclusion must have been true.

In the proof we gave, we had appealed to some familiar mathematical facts, such as the fact that the product of two odd numbers is odd and the fact that every rational number can be expressed as a fraction in lowest terms. These facts, of course, require their own proof, and indeed we shall provide proofs presently.

If $\sqrt{2}$ is not rational, then what kind of number is it? One can view certain developments in the history of mathematics as a process of recognizing problematic issues in our number systems and addressing them with new and better number systems. In ancient days, one had the *whole* numbers $1, 2, 3$, and so on. This number system, however, lacks an additive identity, and by including such an identity, the number zero, we thereby form the *natural numbers* $\mathbb{N}$ with $0, 1, 2, 3$, and so on. Although one can always add two natural numbers together and still have a natural number, this is not true for subtraction, and so we are led to the *integers* $\mathbb{Z}$, which include negative numbers $\ldots, -2, -1, 0, 1, 2, \ldots$ and thereby become closed under subtraction. While one can always multiply two integers and still have an integer, this is not true for division, and so we form the *rational* numbers $\mathbb{Q}$ to accommodate ratios of integers. The fact that $\sqrt{2}$ is irrational shows the need for us to extend our number systems beyond the rational numbers, to the *real numbers* $\mathbb{R}$ and eventually to the *complex numbers* $\mathbb{C}$.

In truth, the actual historical development of the number systems was not like that; it was confused and incoherent. The ancients had the whole numbers and their ratios, and the positive rational numbers, but the number zero was strangely delayed and complicated, appearing early on in some places merely as a placeholder but taken seriously as an actual number only much later. The actual historical development of mathematical ideas is often messy and confounding, riddled with confusion and misunderstanding, yet afterward one can reinvent an imaginary but more logical progression. One can gain mathematical insight from the imaginary history; not every mistake in the actual historical development of mathematics is interesting.

The number $\sqrt{2}$, while irrational, is nevertheless still an *algebraic* number, which means it is the root of a nontrivial integer polynomial, in this case a solution of the equation $x^2 - 2 = 0$. The existence of *transcendental* real numbers, or nonalgebraic numbers, was proved in 1844 by Joseph Liouville. I view Liouville's result as an analogue of the classical $\sqrt{2}$ result, because it shows once again the need for us to extend our number system: the algebraic numbers do not exhaust the numbers we find important. Indeed, we now know that numbers such as e and π are transcendental, and in 1874 Georg Cantor proved in a precise sense that most real numbers are transcendental.

Meanwhile, the complex numbers extend the real numbers with the imaginary numbers, starting with the imaginary unit $i = \sqrt{-1}$. Since the number i solves the equation $x^2 + 1 = 0$, it is an algebraic complex number, but there are also transcendental complex numbers. Mathematicians continue to extend our various number systems, and we now have number systems with various kinds of points at infinity, such as the extended real numbers, which are the real numbers plus $\pm\infty$, as well as other number systems, such as the quaternions, the ordinal numbers, the hyperreal numbers, the surreal numbers, and more.

Returning to the integers and the rationals, let us consider our proof above more closely. We had used some fundamental properties of the integers and the rational numbers, such as the fact that every fraction can be put into lowest terms. That may seem familiar, but can we prove it? Actually, we can completely avoid the issue of lowest terms by arguing as follows.

Slightly revised proof of theorem 1. Suppose toward contradiction that $\sqrt{2}$ is rational. So $\sqrt{2} = p/q$ for some integers p and q, and we may assume that the numerator p is chosen as small as possible for such a representation. It follows as before that $2q^2 = p^2$, and so p^2 and hence also p is even. So $p = 2k$ for some k, which implies that $q^2 = 2k^2$ as before, and so q^2 and hence also q is even. So $q = 2r$ for some r, and consequently $\sqrt{2} = p/q = (2k)/(2r) = k/r$. We have therefore found a rational representation of $\sqrt{2}$ using a smaller numerator, contradicting our earlier assumption. So $\sqrt{2}$ is not rational.   □

This way of arguing, although very similar to the original argument, does not require putting fractions in lowest terms. Furthermore, an essentially similar idea can be used to prove that indeed every fraction can be put in lowest terms.

1.2   Lowest terms

What does it mean for a fraction p/q to be in lowest terms? It means that p and q are *relatively prime*, that is, that they have no common factor, a number $k > 1$ that divides both of them. I find it interesting that the property of being in lowest terms is not a property of the rational number itself but rather a property of the fractional expression used to represent the number. For example, $\frac{3}{6}$ is not in lowest terms and $\frac{1}{2}$ is, yet we say that they are equal: $\frac{3}{6} = \frac{1}{2}$. But how can two things be identical if they have different properties? These two expressions are equal in that they describe the same rational number; the values of the expressions are the same, even though the expressions themselves are different. Thus, we distinguish between the description of a number and the number itself, between our talk about a number and what the number actually is. It is a form of the *use/mention* distinction, the distinction between syntax and semantics at the core of the subject of mathematical logic. How pleasant to see it arise in the familiar elementary topic of lowest terms.

Lemma 2. *Every fraction can be put in lowest terms.*

Proof. Consider any fraction p/q, where p and q are integers and $q \neq 0$. Let p' be the smallest nonnegative integer for which there is an integer q' with $\frac{p}{q} = \frac{p'}{q'}$. That is, we consider a representation $\frac{p'}{q'}$ of the original fraction $\frac{p}{q}$ whose numerator p' is as small as possible. I claim that it follows that p' and q' are relatively prime, since if they had a common factor, we could divide it out and thereby make an instance of a fraction equal to p/q with a smaller numerator. But p' was chosen to be smallest, and so there is no such common factor. Therefore, p'/q' is in lowest terms.   □

This proof and the previous proof of theorem 1 relied on a more fundamental principle, the least-number principle, which asserts that if there is a natural number with a certain property, then there is a smallest such number with that property. In other words, every nonempty set of natural numbers has a least element. This principle is closely connected with the principle of mathematical induction, discussed in chapter 4. For now, let us simply take it as a basic principle that if there is a natural number with a property, then there is a smallest such number with that property.

1.3   A geometric proof

Let us now give a second proof of the irrationality of $\sqrt{2}$, one with geometric character, due to Stanley Tennenbaum. Mathematicians have found dozens of different proofs of this classic result, many of them exhibiting a fundamentally different character from what we saw above.

A geometric proof of theorem 1. If $\sqrt{2}$ is rational p/q, then as before, we see that $p^2 = q^2 + q^2$, which means that some integer square has the same area as two copies of another smaller integer square.

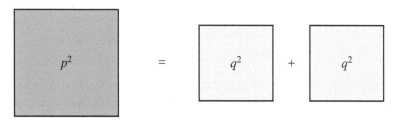

We may choose these squares to have the smallest possible integer sides so as to realize this feature. Let us arrange the two medium squares overlapping inside the larger square, as shown here at the right. Since the large blue square had the same area as the two medium gold squares, it follows that the small orange central square of overlap must be exactly balanced by the two smaller uncovered blue squares in the corners. That is, the area of overlap is exactly

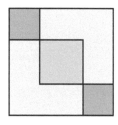

the same as the area of the two uncovered blue corner spaces. Let us pull these smaller squares out of the figure to illustrate this relation as follows.

Notice that the squares in this smaller instance also have integer-length sides, since their lengths arise as differences in the side lengths of previous squares. So we have found a strictly smaller integer square that is the sum of another integer square with itself, contradicting our assumption that the original square was the smallest such instance.   □

1.4   Generalizations to other roots

Let us generalize the theorem in a few ways. For example, an essentially similar argument to our original proof works also for the cube root of 2.

Theorem 3. *The number* $\sqrt[3]{2}$ *is irrational.*

Proof. We can basically follow our first argument for $\sqrt{2}$. Namely, suppose that $\sqrt[3]{2} = p/q$ in lowest terms. By cubing both sides, we conclude that $2q^3 = p^3$. So p^3 is even and therefore p is even, since odd $\times$ odd $\times$ odd is odd. So $p = 2k$ for some k, and therefore $2q^3 = (2k)^3 = 8k^3$. Dividing by 2, we conclude that $q^3 = 4k^3$, and so q^3 is even and therefore q is even, contrary to our assumption that p/q is in lowest terms. $\square$

This argument works essentially the same with $\sqrt[4]{2}$, $\sqrt[5]{2}$, and so on, as the reader will verify in the exercises. Another way to generalize the result is to consider the square roots of numbers other than 2. For example:

Theorem 4. *The number* $\sqrt{3}$ *is irrational.*

Proof. Suppose that $\sqrt{3} = p/q$ in lowest terms. So $3q^2 = p^2$. So p^2 is a multiple of 3. This implies that p is a multiple of 3, since otherwise it would not arise in the prime factorization of p^2. So $p = 3k$ for some integer k; therefore, $3q^2 = p^2 = (3k)^2 = 9k^2$, and so $q^2 = 3k^2$. Thus, q^2 is a multiple of 3, and so q is a multiple of 3. This contradicts our assumption that p/q was in lowest terms. $\square$

This proof relied on the existence and uniqueness of prime factorizations, a familiar fact, which will also be covered in detail in chapter 3. In the exercises, the reader is asked to prove further generalizations. Some generalizations can be handled as a consequence of what we have already done. A *corollary* is a theorem-like statement having an easy proof as a consequence of an earlier theorem.

Corollary 5. *The number* $\sqrt{18}$ *is irrational.*

Proof. Notice that $\sqrt{18} = \sqrt{9 \cdot 2} = 3\sqrt{2}$. If $3\sqrt{2} = p/q$ were rational, then $\sqrt{2} = p/(3q)$, and so $\sqrt{2}$ would also be rational, which it is not. So $\sqrt{18}$ cannot be rational. $\square$

Let us also prove this corollary directly, without relying on the earlier theorem.

Alternative direct proof of corollary 5. Assume toward contradiction that $\sqrt{18}$ is rational. So $\sqrt{18} = p/q$ for some integers p and q, and we may assume that p/q is in lowest terms. Squaring both sides and simplifying, we see that $18q^2 = p^2$, and so p^2 is even. So $p = 2k$ for some integer k; consequently, $18q^2 = p^2 = (2k)^2 = 4k^2$, and so $9q^2 = 2k^2$. So $9q^2$ is even, but since 9 is odd, it must be that q^2 is even and hence also that q is even. So both p and q are even, contrary to our assumption that p/q was in lowest terms. So $\sqrt{18}$ cannot be rational. $\square$

All the theorems and corollaries in this chapter are unified by a grand universal theorem, asserting that $\sqrt[k]{n}$ is irrational unless n is itself a perfect integer kth power, meaning that $n = r^k$ for some integer r. This is equivalent to saying that all the exponents in the prime factorization of n are multiples of k.

Mathematical Habits

State claims explicitly. Do not allow ambiguity in your mathematical claims and theorem statements. Distinguish between similar but inequivalent statements. Formulate your claim to state exactly what you intend.

Know exactly what you are trying to prove. Before embarking on an argument or proof, get completely clear about the meaning and content of the claim that is to be proved.

Insist on proof. Be prepared to prove essentially every mathematical statement you make. When challenged, be prepared to give further and more detailed justification. Do not make assertions that you cannot back up. Instead, make weaker statements that you can prove. Some of the exercises ask you to "prove your answer," but this is redundant, because, of course, you should always prove your answers in mathematical exercises.

Try proof by contradiction. When trying to prove a statement, imagine what it would be like if the statement were false. Write, "Suppose toward contradiction that the statement is false, ..." and then try to derive a contradiction from your assumption. If you succeed, then you will have proved that the statement must be true.

Try to prove a stronger result. Sometimes a difficult or confusing theorem can be proved by aiming at the outset to prove a much stronger result. One overcomes a distracting hypothesis simply by dispensing with it, by realizing it as a distraction, an irrelevant restriction. In the fully general case, one sometimes finds one's way through with a general argument.

> **Generalize.** After proving a statement, seek to prove a more general statement. Weaken the hypothesis or strengthen the conclusion. Apply the idea of the argument in another similar-enough circumstance. Unify our understanding of diverse situations. Seek the essence of a phenomenon.

Exercises

1.1 Prove that the square of any odd number is odd. (Assume that a positive integer is odd if and only if it has the form $2k + 1$.)

1.2 Master several alternative proofs of the irrationality of $\sqrt{2}$, such as those available at www.cut-the-knot.org/proofs/sq_root.shtml, and present one of the proofs to the class.

1.3 Prove that $\sqrt[4]{2}$ is irrational. Give a direct argument, but kindly also deduce it as a corollary of theorem 1.

1.4 Prove that $\sqrt[m]{2}$ is irrational for every integer $m \geq 2$.

1.5 Prove that $\sqrt{5}$ and $\sqrt{7}$ are irrational. Prove that $\sqrt{p}$ is irrational, whenever p is prime.

1.6 Prove that $\sqrt{20}$ is irrational as a corollary of the fact that $\sqrt{5}$ is irrational.

1.7 Prove that $\sqrt{2m}$ is irrational, whenever m is odd.

1.8 In the geometric proof that $\sqrt{2}$ is irrational, what are the side lengths of the smaller squares that arise in the proof? Using those expressions and some elementary algebra, construct a new algebraic proof that $\sqrt{2}$ is irrational. [Hint: Assume that $p^2 = q^2 + q^2$ and that this is the smallest instance in the positive integers. Now consider $2q - p$ and $p - q$.]

1.9 Criticize this "proof." Claim. $\sqrt{n}$ is irrational for every natural number n. Proof. Suppose toward contradiction that $\sqrt{n} = p/q$ in lowest terms. Square both sides to conclude that $nq^2 = p^2$. So p^2 is a multiple of n, and therefore p is a multiple of n. So $p = nk$ for some k. So $nq^2 = (nk)^2 = n^2k^2$, and therefore $q^2 = nk^2$. So q^2 is a multiple of n, and therefore q is a multiple of n, contrary to the assumption that p/q is in lowest terms. □

1.10 Criticize this "proof." Claim. $\sqrt{14}$ is irrational. Proof. We know that $\sqrt{14} = \sqrt{2} \cdot \sqrt{7}$, and we also know that $\sqrt{2}$ and $\sqrt{7}$ are each irrational, since 2 and 7 are prime. Thus, $\sqrt{14}$ is the product of two irrational numbers and therefore irrational. □

1.11 For which natural numbers n is $\sqrt{n}$ irrational? Prove your answer. [Hint: Consider the prime factorization of n, and consider especially the exponents of the primes in that prime factorization.]

1.12 Prove the unifying theorem mentioned at the end the chapter, namely, that $\sqrt[k]{n}$ is irrational unless n is itself an integer kth power.

1.13 Prove that the irrational real numbers are exactly those real numbers that are a different distance from every rational number. Is it also true if you swap "rational" and "irrational"?

2 Multiple Proofs

In mathematics we often have more than one proof of a given theorem. In the previous chapter, for example, we gave several proofs that $\sqrt{2}$ is irrational. I mentioned this to my daughter (ten years old), and she replied, "Why would you ever need more than one proof? If you've proved it once, then you know it is true." She is surely correct about that—one correct proof is sufficient for us to know that a theorem is true.

Nevertheless, we often find it valuable to have multiple proofs of a theorem, especially in the case of an important or central result or in a case where the proofs are extremely different. Why is this? Is it because we doubt the correctness of our proofs and seek to buttress our knowledge against error? That would make sense, and surely our confidence naturally increases with multiple proofs, but ultimately, I do not believe that this is the main explanation.

Rather, as I see it, the main reason we value multiple proofs of a theorem is that different arguments, especially when they are extremely different and highlight different fundamental aspects of a topic, deepen our mathematical understanding and appreciation of a mathematical phenomenon.

Having a proof of a theorem, after all, is not merely about establishing the truth of that theorem statement but, rather, about explaining why the theorem is true, about giving us the conceptual framework in which to understand the mathematical fact more deeply. The proof, sometimes much more so than the theorem statement itself, reveals the hidden structure of mathematical reality, a wealth of mathematical meaning at the core, elucidating deep connections between mathematical phenomena. One gains mathematical insight not just by knowing the fact but by pondering and internalizing the proof, the explanation, and meaning behind the fact. Furthermore, different mathematical arguments can often suggest different avenues of generalization, where we might hope to allow similar reasoning in similar-enough circumstances.

2.1 $n^2 - n$ is even

I should like in a small way to illustrate this latter phenomenon about generalizations with a toy example, which I learned from Benjamin Dickman. Consider the following elementary theorem.

Theorem 6. *For any natural number n, the number $n^2 - n$ is even.*

The theorem is not difficult to prove, and we shall provide several proofs, from different perspectives, which can be described under the following rubrics:

- Proof by cases
- Proof by high-school algebra
- Proof by induction
- Proof by geometrical figure
- Proof by combinatorics
- Proof by modular arithmetic
- Proof using the identity $1 + 2 + \cdots + (n - 1) = n(n - 1)/2$

Please try to prove the theorem on your own, using one these rubrics or your own idea. Select a proof idea that you think you can carry out, close the book, and prove the theorem now.

Interlude...

Welcome back. Did your proof succeed?

2.2   One theorem, seven proofs

Let us give proofs of the theorem following each of the perspectives mentioned above. We begin with the method of proof by cases.

Proof #1 (by cases). We consider two cases, depending on whether n is even or odd. If n is even, then n^2 is also even (why?), and so $n^2 - n$ is the difference of two even numbers, which is even (why?). If n is odd, then n^2 is odd (why?), and so $n^2 - n$ is the difference of two odd numbers, which is even (why?). So in any case, $n^2 - n$ is an even number, as the theorem claims. (The reader shall answer the *why* questions in exercise 2.1.)   □

Next, consider a proof using some elementary algebra.

Proof #2 (by high-school algebra). We may factor the expression as $n^2 - n = n(n - 1)$, and so we see that $n^2 - n$ is the product of two consecutive numbers. At least one of those must be even, and so $n^2 - n$ is even.   □

And now a proof by induction. Mathematical induction is a method of proving that something is true for all natural numbers by showing that it starts out true at $n = 0$ and that, furthermore, truth is preserved from each number to the next. We shall cover this method in detail in chapter 4, but for now let us simply use the method.

Proof #3 (by induction). The claim that $n^2 - n$ is even is true when $n = 0$, since 0 is even. So it starts out true. To see that the fact is propagated from each number to the next, suppose that $n^2 - n$ is even and consider the next instance, $(n + 1)^2 - (n + 1)$. By using some elementary algebra, this is equal to $n^2 + 2n + 1 - n - 1 = n^2 + n$, which can be written as $(n^2 - n) + 2n$. Thus, in moving from $n^2 - n$ to $(n + 1)^2 - n^2$ we have added $2n$, an even number. And so if the statement is true at a number n, then it remains true at $n + 1$, and so the truth of the statement is propagated from each number to the next. So by mathematical induction, it is true for all numbers. □

Next, we prove the theorem using geometry.

Proof #4 (by geometrical figure). Consider an $n \times n$ square, consisting of n^2 small squares. If we remove the diagonal, which has n squares on it, there are exactly $n^2 - n$ squares remaining.

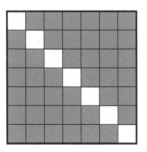

Notice that the figure is completely symmetric, and every small square on one side of the diagonal can be paired with a corresponding square on the other side of the diagonal. So $n^2 - n$ is even. □

We now prove the theorem using ideas from combinatorics.

Proof #5 (by combinatorics). The reader may happen to know that the number of ways to choose 2 things from a set of $n - 1$ things is $n(n - 1)/2$, and we shall prove such facts in chapter 5. Since this number is an integer, it follows that the numerator $n(n - 1)$ must be a multiple of 2, and so we conclude that $n^2 - n$ is even. □

And next we prove the theorem using modular arithmetic.

Proof #6 (by modular arithmetic). Modular arithmetic, modulo 2, refers to the system of arithmetic system that arises by replacing every integer with its residue modulo 2, or in other words, with its remainder after dividing by 2. (More generally, one can carry out modular arithmetic modulo any given base $b > 1$.) The main point of modular arithmetic is that the arithmetic operations are indeed well defined on these residues. In modular arithmetic modulo 2, residues are either 0 or 1, and so $n^2 - n$ mod 2 is either $0^2 - 0 = 0$ or $1^2 - 1 = 0$, and in either case we get 0 mod 2, which means $n^2 - n$ is even. □

Finally, we prove the theorem using the following summation identity, which the reader may happen to know:
$$1 + 2 + \cdots + (n - 1) = n(n - 1)/2.$$

This identity is important in the subject of finite combinatorics, since as we mentioned this quantity is the number of ways of choosing two objects from a set of n objects. We shall prove it in chapter 4 and again with a different proof in chapter 6.

Proof #7 (using summation identity). The point is that identity immediately implies that $n(n-1)/2$ is an integer, since the left side of the identity is an integer. And so the numerator $n(n - 1)$ must be even, and therefore $n^2 - n$ is even. □

Can you find any other proofs of theorem 6?

2.3   Different proofs suggest different generalizations

We had mentioned that mathematicians sometimes find it valuable to have multiple proofs of a theorem, because the different proof ideas may emphasize different aspects of the theorem, which may suggest different avenues for generalization. Even slightly different but equivalent ways of stating a theorem could suggest different generalizations. How does that play out here? Several of the arguments turned on the fact that $n^2 - n = n(n - 1)$, and so the two assertions

2 divides $n(n - 1)$,
2 divides $n^2 - n$

are simply equivalent to each other. But perhaps these different ways of stating the result already suggest different ways to generalize the statement.

For example, let us try to generalize the theorem by considering multiples of 3, rather than 2. The two statements above perhaps suggest the following generalizations:

3 divides $(n + 1)n(n - 1)$, and
3 divides $n^3 - n$.

These statements are actually equivalent to each other, as one can see by verifying that $(n + 1)n(n - 1) = n(n^2 - 1) = n^3 - n$. Furthermore, the high-school-algebra proof above generalizes to show that indeed the statements are true, since $(n + 1)n(n - 1)$ is the product of three consecutive numbers, one of which must be a multiple of 3, and so the product is indeed a multiple of 3.

How about multiples of 4? The high-school proof again generalizes to show that the product of any four consecutive numbers is a multiple of 4 (and in fact these products are all multiples of 24, which the reader will prove in exercise 2.7). But meanwhile, $2^4 - 2$ equals 14, which is not a multiple of 4. So the analogues of the two statements are no longer equivalent in the case where we consider multiples of 4. This is now a very interesting situation, since we are led to consider the question: for which numbers k is $n^k - n$ always a multiple of k? This question leads into the deeper waters of number theory.

Meanwhile, the geometric proof suggests the idea of going to three dimensions. For example, we can see that $n^3 - n^2$ is always even, since you can cut a cube, like a cake, by laying the knife along the diagonal at the top and cutting straight down, removing the n^2 many cubelets appearing on that slice. What is left has two equal halves, and so $n^3 - n^2$ is even. Of course, we could also have observed simply that $n^3 - n^2 = n(n^2 - n)$, which is therefore a multiple of an even number and hence even.

Mathematical Habits

Try it yourself first. When encountering a new theorem while reading, close the book or article and try first to prove it yourself. Push your argument through as far as you can. When you get stuck, consult the original proof for a clue about a concept or perspective you had missed, and armed with that clue, close the book or article and try once again to prove the theorem on your own. This way of reading will give you a greater understanding of the challenges and a greater appreciation for the value of a key step or idea, which will also enable you to learn the argument very deeply. Students who read proofs this way will inevitably become strong mathematicians.

Seek multiple proofs. Strive to prove your statements with arguments arising from totally different perspectives on a problem, which might give insight on different aspects of the situation.

Prove a special case. Faced with a difficult or stubborn proof, make an additional assumption and prove the statement in that special case.

> **Try proof by cases.** After proving one special case, look for the remaining cases. Sometimes one can prove a statement by totally separate arguments in the various possible cases.

Exercises

2.1   Prove that the sum, difference, and product of two even numbers is even. Similarly, prove that the sum and difference of two odd numbers is even, but the product of odd numbers is odd.

2.2   True or false: if the sum of one pair of positive integers is larger than the sum of another pair, then the product also is larger.

2.3   True or false: if the product of one pair of positive integers is larger than the product of another pair, then the sum also is larger.

2.4   Prove that the number of ways to choose two things from a set of n things is $n(n-1)/2$. In other words, $\binom{n}{2} = n(n-1)/2$.

2.5   Prove that the product of k consecutive integers is always a multiple of k.

2.6   Prove that the product of any three consecutive integers is a multiple of 6. Conclude that $n^3 - n$ is a multiple of 6 for every integer n.

2.7   Prove that the product of any four consecutive positive integers is a multiple of 24.

2.8   Generalize exercise 2.7 to the product of five consecutive numbers, and to six, and so on. Formulate a conjecture concerning the product of any n consecutive positive integers being a multiple of some number depending on n. Can you prove it?

2.9   State and prove a theorem concerning positive integers k for which the product of any $k-1$ consecutive positive integers is a multiple of k.

2.10 Can you criticize the following argument? Claim. $n^3 - n$ is always even for any natural number n. Proof. Consider an $n \times n \times n$ cube, consisting of n^3 many small cubes. Remove the n small cubes on the long diagonal through the cube, so there are $n^3 - n$ many cubes remaining. Since the cube is separated into two identical and symmetric pieces by the main diagonal, one on each side, it must be that $n^3 - n$ is even. $\square$

Credits

The essential idea of this chapter was adapted from a treatment of Benjamin Dickman, who first proposed the $n^2 - n$ example on MathOverflow, Dickman (2013), and later developed it more fully in his article, Dickman (2017), written for mathematics educators.

3 Number Theory

Number theory is celebrated by mathematicians as a pure form of abstract thought, a distillation of reason. Carl Friedrich Gauss called it the "queen of mathematics," while G. H. Hardy, in *A Mathematician's Apology*, admired its pure, abstract isolation, praising the fact that it was unencumbered by the physical world, without use or application.

And yet, despite this, in a strange and surprising twist of fate, the theory has in contemporary times found key practical applications; deep number-theoretic ideas, for example, lie at the core of cryptography and internet security. Our dreamy iridescent theory of numbers, it turns out, supposedly without use or application, does in fact have important applications, vital for commerce and communication, so much so that number-theoretic ideas are now firmly established via cryptography in the foundations of our economy.

3.1 Prime numbers

Let us develop some elementary number theory, beginning with the prime numbers. What is a prime number? At this question, I imagine, perhaps a nervous laugh goes through the classroom—of course we are all familiar with prime numbers, right? A helpful student suggests, tentatively, that a number is prime if the only divisors of the number are 1 and itself. Is this a good definition? Let us quibble. One minor issue is that numbers can have negative divisors. We want 3 to be prime, to be sure, but does the factorization $3 = (-3) \cdot (-1)$ mean that it has -3 and -1 also as divisors? Let us therefore interpret the student's proposal to refer only to *positive* divisors. Similarly, we do not want $\frac{7}{2} \cdot 2$ to rule out 7 as prime, and so we should interpret the student's proposal as referring only to positive *integer* divisors.

A more serious issue, however, is the question of whether the number 1 should count as prime. Is 1 prime? Since 1 has no positive integer divisors other than 1 and itself, it would seem that, according to the student's proposal, we would say that, indeed, 1 is prime. Nevertheless, this would go against the advice of many mathematicians, who have come to the conclusion that we do not actually want to include 1 amongst the prime numbers. On this view, the student's suggestion is not quite exactly right.

I should like to emphasize that, as mathematicians, we are free to define our terms to mean whatever we want. We could define our notion of prime number so that 1 turns out to be prime, or we could have defined the notion differently, so that 1 would not count as prime. There is no issue here of *discovering* whether or not the number 1 is prime; we can define our technical terms as we wish. Rather, what is at stake here is what we want the word *prime* to mean and whether we want to use this word in a way that makes the number 1 prime or makes it not prime.

All things considered, mathematicians have concluded that it is more convenient in many contexts if we say that 1 is *not* prime. To give one example, if the number 1 counted as prime, then it would no longer be true that every positive integer had a unique prime factorization, as in $12 = 3 \cdot 2^2$, since we could always add more 1s the factorization, like this: $12 = 3 \cdot 2^2 \cdot 1 \cdot 1 \cdot 1$. If 1 counted as prime, we would find ourselves often having to add exceptions to many theorems about prime numbers, to say in effect, "except for 1." Ultimately, it is more convenient simply to say that 1 does not count as prime. So let us make our official definition as follows:

Definition 7. An integer p is *prime* if $p > 1$ and the only positive integer divisors of p are 1 and p.

Students of abstract algebra might recognize that this definition is actually not concerned with primality but with *irreducibility*. Specifically, in the algebraic structures known as rings, an element is defined to be irreducible if it admits no nontrivial factorization, as in the definition above. An element p of a ring is *prime*, in contrast, if it is nonzero, it has no multiplicative inverse, and whenever p divides a product ab, then either p divides a or p divides b. Although in the class of rings known as integral domains every prime element is irreducible, the converse is not always true, and many rings and integral domains have irreducible elements that are not prime. Nevertheless, it will follow from lemma 13 below that in the positive integers the prime numbers are the same as the irreducible numbers, and so it will be fine for us to stick with the traditional definition above as our official definition of what it means to be prime.

3.2 The fundamental theorem of arithmetic

The reader has likely had experience factoring numbers as a product of prime numbers. Twelve factors as $12 = 3 \cdot 2^2$, and the number 8470 can be factored first as the product $10 \cdot 847$, and then further broken into primes as $2 \cdot 5 \cdot 7 \cdot 11 \cdot 11$. Let me ask an innocent question:

Question 8. For a given number, will we always get the same prime factorization?

Of course, I mean the *same* here in the sense that $3^2 \cdot 5 \cdot 7$ counts as the same factorization as $5 \cdot 3 \cdot 7 \cdot 3$—we have simply rearranged the same prime factors. Are prime factorizations always unique in this sense? Perhaps. Can we imagine that, for extremely large numbers, the outcome of the prime factorization might depend on how we had proceeded to compute it? Perhaps. If we had started with a slightly different product at the first step, might we have found ourselves ultimately with a different prime factorization in the end?

The number 11543, for example, can be factored as $7 \cdot 17 \cdot 97$, and these factors are each prime. Do we know that there is not also some other way to factor it into primes? For a comparatively small number like this, we might hope to try out all the other candidates and be convinced this way. That method is actually impractical, even for numbers such as 11543, and for much larger numbers like 6534653454534354634563456534655534354652, it is downright infeasible. Do we actually know that this number has a unique prime factorization?

When factoring numbers, to be sure, we routinely refer to *the* prime factorization. But perhaps we should be saying merely that we have *a* prime factorization? Can there be more than one? On my view, this is a serious question, more difficult than it might initially appear. The fact of the matter is that the usual naive treatment of prime factorization simply does not touch on the question of uniqueness. We have become familiar with the uniqueness of prime factorizations largely by observing many instances of prime factorization, without ever encountering a situation where a number admits more than one factorization. Does this experience constitute proof? No, of course not.

Meanwhile, prime factorizations are indeed unique, and this fact is the first deep theorem of number theory, known as the fundamental theorem of arithmetic:

Theorem 9 (Fundamental theorem of arithmetic). *Every positive integer can be expressed as a product of primes, and furthermore, this factorization is unique up to a rearrangement of these prime factors.*

We shall prove this theorem presently, but before doing so, let me remark on a certain issue arising in the statement of this theorem and a habit that mathematicians have. Namely, the theorem says that every positive integer is a product of primes; but is this right? What about the number 5? Yes, it is prime, but is it a *product* of primes?

Mathematicians will say yes, 5 is a product of primes, a product consisting of just one factor, the number 5 itself. You might reply, that is not a product at all! Should not the theorem say, "Every positive integer is either prime or a product of primes"? Mathematicians will insist, nevertheless, that it is a good idea to consider 5 and the other prime numbers as products of primes, degenerate products if you will, products consisting of just one factor. The reason is that our mathematical theories often become more robust when we incorporate the trivial or degenerate instances of a phenomenon into the fundamental definitions. Every square counts also as a rectangle; every equilateral triangle is also isosceles. This practice often leads to a smoother mathematical analysis in the end. A rectangle is precisely a quadrilateral with four right angles, for example, but this would not be true if we did not count squares also as rectangles. In the same way, we allow ourselves to refer to the product of just one number, without being multiplied by any other number.

One sometimes hears that 5^n means that we multiply 5 by itself n times. This is not really accurate, however, if what is meant is that we have n multiplications, since 5^2 means 5×5, which is only one act of multiplication; we would be more correct, therefore, to say that n refers to the number of factors, rather than to the number of times we are multiplying. This is a *fence-post* error, discussed further in chapter 5. So we regard 5^1 as a product consisting of just one factor. Similarly, we take $5^0 = 1$ as a product with no factors at all, the empty product. This is the sense in which 1 is a "product" of prime numbers.

The thing to notice here is that, even if we had stated the theorem as, "Every positive integer is either prime or a product of primes," it still would not be correct without this empty product convention, since the number 1 is a positive integer, but it cannot be expressed as a product of primes except as the empty product. Without the empty and singleton product considerations, therefore, we would have to say, "Every integer greater than 1 is either prime or a product of primes." But is it not both simpler and more elegant to state the theorem as we have in theorem 9? We have included the primes themselves each as a product with only one factor and the number 1 as the empty product.

So let us now prove the fundamental theorem, which makes two essentially different claims: an existence claim and a uniqueness claim. Namely, the existence claim is that every positive integer admits at least one prime factorization, and the uniqueness claim is that every number admits at most one prime factorization, in the sense that any two factorizations of the same number are rearrangements of one another. Let us begin with the existence claim, since this is perhaps more familiar, as well as easier to prove.

Theorem 10 (Fundamental theorem, existence). *Every positive integer can be expressed as a product of prime numbers.*

Proof. Let us prove that every positive integer has the property. The number 1 is expressed as the empty product, and so we have made a start. Suppose that all the numbers up to some number n have the property, where $n > 1$, and now consider whether n itself has the property. If n happens to be prime, then indeed, it is expressible as a product of just one prime factor, itself, and so it would have the property. Otherwise, n is not prime, and so we may factor it as $n = ab$ for some numbers a and b, both smaller than n. Because we assumed that the property holds up to n, we know that both a and b are expressible as products of primes. So $a = p_1 \cdots p_k$ and $b = q_1 \cdots q_r$ for primes p_i and q_j, where $1 \le i \le k$ and $1 \le j \le r$. By simply combining these products, we can now realize n also as a product of primes:

$$n = ab = p_1 \cdots p_k \cdot q_1 \cdots q_r.$$

So the property does indeed hold at n. We have therefore proved that whenever the property holds up to a number n, then it holds at the number n. In other words, there can be no minimal counterexample to the property; thus, there can be no counterexample at all. So every number has the property. $\square$

We used the method of *minimal counterexamples*, by which one shows that a property holds of all positive integers by showing that there can be no smallest counterexample to the property, and hence no counterexample at all. This method is closely related to (and essentially identical to) the method of *mathematical induction*, which we shall explore more fully in chapter 4.

Some mathematicians may have preferred to cover the method of induction before proving the fundamental theorem of arithmetic, but my preference was to mount these simple minimal-counterexample arguments in this chapter, using them in part as an introduction to inductive reasoning. In chapter 4, we shall give a more thorough general account of the theory of mathematical induction.

3.3   Euclidean division algorithm

In order to prove the uniqueness part of the fundamental theorem of arithmetic, we shall rely on some classic elementary number theory, which we now develop. Let us begin with the familiar fundamental principle that we can always divide integers, possibly with remainder, in such a way that the remainder is less than the number by which we are dividing. This fact is known as the Euclidean division algorithm. This terminology, quite old and firmly established, perhaps conflates the mathematical fact that the quotient and remainder exist with the algorithms or procedures that one might use to find them.

Lemma 11 (Euclidean division algorithm). *For any two positive integers n and d, there are unique integers q and r for which $n = qd + r$ and $0 \leq r < d$.*

Note how we have chosen variable names so as to aid our understanding, since the lemma is fundamentally concerned with the operation of dividing n by d, so n stands for *numerator*, d for *denominator* or *divisor*, q for *quotient*, and r for *remainder*.

Proof. To prove uniqueness, suppose that $qd + r = q'd + r'$, where $0 \leq r, r' < d$. It follows that $r - r' = (q' - q)d$, and so $r - r'$ is a multiple of d. Since also $|r - r'| < d$, it follows that $r - r' = 0$ and so $r = r'$, which implies $q = q'$. So the representation is unique when it exists. For existence, we shall prove that there can be no smallest failing instance of the lemma. Specifically, we shall prove that if the lemma holds up to a number n, then it also holds at n. So suppose that n and d are numbers and that the statement of the lemma holds with d and any number smaller than n. If $n < d$, then we can write $n = 0 \cdot d + n$, and this fulfills the requirement that $0 \leq r < d$. Similarly, if $n = d$, then we may write $n = 1 \cdot d + 0$, which also verifies the desired property. So we may assume that $d < n$. In this case, $n - d$ is a positive integer smaller than n. By the assumption on n, the lemma holds for d and $n - d$, and so there are numbers q and r with $n - d = qd + r$ and $0 \leq r < d$. By adding d to both sides, we see that $n = (q + 1)d + r$, which fulfills the desired statement for n and d. So there can be no minimal counterexample to the lemma, and consequently, there can be no counterexample at all. So the lemma holds for all n and d. □

Next, we prove Bézout's identity. Recall from chapter 1 that integers are *relatively prime* if they have no common factor larger than 1.

Lemma 12 (Bézout's identity). *If integers a and b are relatively prime, then there are integers x and y for which* $1 = ax + by$.

Proof. Assume that integers a and b are relatively prime. Let d be the smallest positive integer that is expressible as an *integer linear combination* of a and b, that is, as $d = ax + by$ for some choice of integers x and y. Certainly $d \leq a$, since we can write $a = a \cdot 1 + b \cdot 0$. I claim that d divides both a and b. To see this, apply the Euclidean algorithm to express $a = kd + r$ for some integer k and remainder r, with $0 \leq r < d$.

Putting our equations together, observe that

$$r = a - kd = a - k(ax + by) = (1 - kx)a + (-ky)b.$$

We have therefore expressed r as an integer linear combination of a and b. Since $r < d$ and d was the smallest positive such combination, it follows that r must be 0. In other words, $a = kd$ is a multiple of d, as claimed. A similar argument shows that b also is a multiple of d, and so d is a common factor of a and b. Since these numbers are relatively prime, it must be that $d = 1$, and so we have achieved $1 = ax + by$, as desired.   □

Lemma 13 (Euclid's lemma). *If p is prime and p divides ab in the integers, then p divides a or p divides b.*

Proof. Assume that p is prime and that p divides ab. If p does not divide a, then a and p must be relatively prime, since there are no other nontrivial factors of p. By Bézout's lemma (lemma 12), it follows that $1 = ax + py$ for some integers x and y. Multiplying both sides by b, we see that

$$b = abx + pby.$$

Since p divides ab, it follows that p divides the right-hand side of this equation, and so p divides b. So we have proved that if p does not divide a, then it divides b. And so p must divide one of them.   □

We can generalize lemma 13 to the situation of many primes:

Lemma 14. *If a prime p divides a product of integers* $n_1 n_2 \cdots n_k$, *then p divides some* n_i.

Proof. We know by lemma 13 that this lemma is true when there are only two factors. Suppose that this lemma holds when there are fewer than k factors, and that we have a prime number p that divides a product $n_1 \cdot n_2 \cdots n_k$ with k factors. The trick is to look upon the product $n_1 n_2 \cdots n_k$ as a product of just two things, like this: $n_1 \cdot (n_2 \cdots n_k)$. Since p divides this product of two things, we may conclude by lemma 13 that either p divides n_1 or p divides the rest of the product $n_2 \cdots n_k$. In the first case, we are done immediately, and in the second case, since there are now fewer than k factors in the product, we conclude by our assumption on k that p must divide one of the n_i for $2 \le i \le k$. So in any case, p divides some n_i, and the lemma is proved. □

3.4   Fundamental theorem of arithmetic, uniqueness

Finally, we can prove the uniqueness part of the fundamental theorem of arithmetic.

Theorem 15 (Fundamental theorem, uniqueness). *Every positive integer has at most one prime factorization, in the sense that any two factorizations are simply rearranging the order of the prime factors appearing in them.*

Proof. We have already proved the existence claim in theorem 10. What remains is the uniqueness claim. Suppose that all the numbers smaller than a number n have at most one representation as a product of primes (unique up to rearranging the order in which the prime factors appear in the product). Suppose that $n = p_1 \cdots p_k = q_1 \cdots q_r$ are two representations of n as a product of primes. Since p_1 divides n, it follows by lemma 14 that p_1 must divide one of the q_j, and since these are all prime, it must be equal to one of the q_j. It might as well be q_1, by rearranging the qs. But in this case, we have $p_2 \cdots p_k = q_2 \cdots q_r$, since these are both n/p_1, and by our assumption on n, these two products are a simple rearrangement of each other. So the original products also are obtained by rearranging, and we are done. □

The fundamental theorem of arithmetic (theorem 9) amounts to the combination of the existence claim of theorem 10 and the uniqueness claim of theorem 15, so it is now proved.

3.5   Infinitely many primes

Let us turn now to another classical result, the fact that there are infinitely many primes. This is a classic argument, often attributed to Euclid, known for thousands of years.

Theorem 16. *There are infinitely many prime numbers.*

Proof. Suppose that you have a list of finitely many prime numbers:

$$p_1, \ p_2, \ \ldots, \ p_n.$$

Let $N = (p_1 p_2 \cdots p_n) + 1$, the result of multiplying them together and adding 1. Observe that if you should divide N by any particular prime number p_i on your list, then there will be a remainder of 1. In particular, this number N is not divisible by any prime number on your list. But N is a product of primes, as every natural number is, and so there must be a prime that is not on the list. Thus, no finite list of numbers includes all the primes, and so there must be infinitely many of them. □

Sometimes one sees this argument given as a proof by contradiction, like this:

Proof. Assume toward contradiction that there are only finitely many primes, p_1, p_2, ..., p_n. Multiply them all together and add one $N = p_1 p_2 \cdots p_n + 1$. This new number N is not a multiple of any p_i, and so its prime factorization must involve new primes, not on the list, a contradiction. □

Is it better to give a proof by contradiction or directly? This second proof seems perfectly fine. Euclid had not used proof by contradiction but, rather, proved that every finite list of primes can be extended. Mathematicians often prefer direct proofs over proofs by contradiction. One reason, to my way of thinking, is that direct proofs usually carry more information about the mathematical background—they paint a fuller picture of mathematical reality. When one proves an implication $p \implies q$ directly, one assumes p and derives further consequences p_1, p_2, and so on, before ultimately concluding q. Thus, one has learned a whole context about what it is like in the p worlds. Similarly, with a proof by contraposition, one assumes the negation of the conclusion $\neg q$ and derives consequences about what it is like in the worlds without q, before finally concluding $\neg p$, negating the hypothesis. But in a proof by contradiction, in contrast, one assumes something that ultimately does not hold in any world; the argument often seems to tell us little beyond the brute fact of the implication $p \implies q$ itself.

Next, let us prove that although there are infinitely many prime numbers, nevertheless there are also long stretches of numbers with no prime numbers at all.

Theorem 17. *There are arbitrarily large intervals in the positive integers containing no prime numbers.*

Proof. Consider any positive integer $n \geq 2$. Notice that $n! + k$ is a multiple of k, whenever $1 \leq k \leq n$, since in this case it is the sum of two multiples of k. By considering $2 \leq k \leq n$, we have therefore found an interval of positive integers, from $n! + 2$ up to $n! + n$, containing no prime numbers. This interval (inclusive) has $n - 1$ numbers in it, and so there are arbitrarily large intervals of nonprimes in the positive integers. □

It is natural to inquire about the density of the prime numbers. Let $\pi(n)$ be the *prime-counting* function, the number of prime numbers $p \leq n$.

| 2 | 521 | 1117 | 2003 | 3119 | 4271 | 4999 |

The figure above illustrates the density of primes, with a narrow blue line for each prime number up to five thousand, spaced out on the number line (some selected primes are indicated in red). The pattern is irregular, with some intervals having primes clumped near each other and other intervals more sparsely populated, with perhaps a barely perceptible decay in the density as one moves to higher numbers. Since theorem 17 shows that there are large primeless intervals, we should expect to find increasing white patches, without any blue lines, if the figure were continued to larger numbers.

In contrast to the previous result, let us now provide a lower bound on the number of prime numbers. Define that a positive integer r is *square-free* if it is not a multiple of any square number bigger than 1; equivalently, by exercise 3.12, a positive integer is square-free if all the primes in its prime factorization have exponent 1.

Theorem 18 (Erdős). *The prime-counting function π has a lower bound provided by the inequality*

$$\frac{\log_2(n)}{2} \leq \pi(n).$$

Proof. Every natural number can be factored as rs^2, where r is square-free, as the reader will verify in exercise 3.13. How many square-free numbers are there less than n? Well, every square-free number is a product of distinct primes, and so it is determined by the set of primes that divide it. So the number of square-free numbers up to n will be at most $2^{\pi(n)}$, which is the number of sets of primes up to n. Second, the number of squares up to n is at most $\sqrt{n}$, since if $s^2 \leq n$ then $s \leq \sqrt{n}$, and s^2 is determined by s. So the number of numbers up to n having the form rs^2, where r is square-free, is at most $2^{\pi(n)} \sqrt{n}$. Since there are n positive integers less than or equal to n, we conclude that

$$n \leq 2^{\pi(n)} \sqrt{n}.$$

Dividing both sides by $\sqrt{n}$ leads to $\sqrt{n} \leq 2^{\pi(n)}$, and then taking the logarithm (base 2) enables us to conclude that

$$\log_2(\sqrt{n}) \quad \leq \quad \log_2(2^{\pi(n)}),$$

which amounts to the inequality

$$\tfrac{1}{2} \log_2(n) \quad \leq \quad \pi(n),$$

as desired. □

An *arithmetic progression* is a sequence of numbers of the form p, $p + e$, $p + 2e$, $p + 3e$, and so on. Two numbers are *relatively prime* if they have no nontrivial common factor.

Theorem 19. *There are arbitrarily long arithmetic progressions consisting of relatively prime numbers.*

Proof. Fix any number d, and let $e = d!$, the factorial of it. Let p be a prime number larger than e, and consider the numbers

$$p \qquad p + e \qquad p + 2e \qquad p + 3e \qquad \cdots \qquad p + (d-1)e,$$

in other words, the numbers of the form $p + ei$, where $i < d$. These form an arithmetic progression of length d and period e. Let me show that these numbers are all relatively prime. Suppose that q is a prime factor of two of the numbers, $p + ei$ and $p + ej$, where $i < j < d$. It follows that q divides the difference of the two numbers, which is equal to $e(j - i)$. Since i and j are both less than d and $e = d!$, it follows that all of the prime factors of $e(j - i)$ are at most d, so $q \le d$. In this case, q divides e and hence ei, and since it also divides $p + ei$, it must be that q divides p, which is prime. So $q = p$, contrary to our choice of p to be larger than e. $\qquad\square$

It had been a long-standing open question, since at least 1770, whether one could find arbitrarily long arithmetic progressions of prime numbers (not merely relatively prime). This much harder question was finally settled in the affirmative in a celebrated 2004 result of Ben Green and Terence Tao. The Green-Tao theorem states that for every natural number d, there is an arithmetic progression of length d consisting entirely of prime numbers.

Mathematical Habits

Choose variable names well. Choose sensible variable names that help remind you of their meaning. But also try to follow established variable-naming conventions, when possible, in order to hook into your readers' expectations about what kind of quantity your variable represents. For example, many mathematicians use variables n and m to represent natural numbers or integers, while p and q are probably prime numbers, the variables x and y frequently represent real numbers, z is often a complex number, and f and g are likely functions. Conventions can vary between particular mathematical specializations.

Use technical words with accuracy and precision. Recognize that mathematical words often carry extremely precise meanings, far more explicit and detailed than the meanings conveyed in ordinary language by those words. Use technical vocabulary strictly to carry these more specific meanings.

Define your technical terms. Provide explicit formal definitions for your mathematical terms, and use the words in accordance with their definitions. Clarify undefined terms; do not allow vagueness and ambiguity to slip into your analysis simply because you have not taken the trouble to define your terms.

Try proving implications directly. When proving a statement of the form, "if p, then q," try starting your argument by writing, "Assume p," and then argue for q. This method is called a direct proof of the implication.

Try proving implications via the contrapositive. When proving a statement of the form "if p, then q," consider the contrapositive, "if not q, then not p." This is logically equivalent to the original statement and sometimes admits a smoother argument, particularly when q has a negative form, in which case "not q" becomes a positive assumption. Write, "To prove the contrapositive, assume not q." And then argue for not p.

Exercises

3.1 Prove that a positive integer is prime if and only if it has exactly two positive integer divisors.

3.2 Show that a positive integer $p > 1$ is prime if and only if, whenever p divides a product ab of integers, then either p divides a or p divides b.

3.3 Prove a stronger version of Bézout's lemma, namely, that for any two integers a and b, the smallest positive number d expressible as an integer linear combination of a and b is precisely the greatest common divisor of a and b.

3.4 Read Timothy Gower's blog post, "*Why isn't the fundamental theorem of arithmetic obvious?*" (https://gowers.wordpress.com/2011/11/13/w) and his follow-up post, "*Proving the fundamental theorem of arithmetic*" (https://gowers.wordpress.com/2011/11/18/p). Write a summary.

3.5 Show that if we count the number 1 as prime, then the uniqueness claim of the fundamental theorem of arithmetic would be false. (And this is one good reason not to count 1 as amongst the prime numbers.)

3.6 Prove that if one prime divides another, then they are equal. (This was used in the proof of theorem 9.)

3.7 Mathematician Evelyn Lamb fondly notes of any large prime number presented to her that it is "one away from a multiple of 3!" And part of her point is that this is true whether you interpret the exclamation point as an exclamation or instead as a mathematical sign for the factorial. Prove that every prime larger than 3 is one away from a multiple of 3!. [Hint: Consider the remainder of the number modulo 6.]

3.8 Prove that if a, b, and c are positive integers and the product abc is a multiple of 6, then one of the numbers ab, ac, or bc is a multiple of 6.

3.9 Define that an integer is *even* if it is a multiple of 2; and otherwise it is *odd*. Show that every odd number has the form $2k+1$ for some integer k. Conclude that any two consecutive integers consist of one even number and one odd number.

3.10 Prove or refute the following: For any list of primes $p_1, \ldots, p_n$, the number $(p_1 p_2 \cdots p_n) + 1$ is prime.

3.11 Criticize the following "proof." Claim. There are infinitely many primes. Proof. Consider any number n. Let $n!$ be the factorial of n, and consider the number $p = n! + 1$. So p is larger than n, and it has a remainder of 1 dividing by any number up to n. So p is prime, and we have therefore found a prime number above n. So there are infinitely many prime numbers. $\square$

3.12 Prove that a positive integer is square-free if and only if all of the exponents in its prime factorization are 1.

3.13 Prove that every positive integer m can be factored as $m = rs^2$, where r is square-free. [Hint: Let r be the product of the primes having odd exponent in the prime factorization of m. Alternatively, let s be the largest number such that s^2 divides m, and argue that the quotient is square-free.]

3.14 Prove or refute the following: The lower bound on $\pi(n)$ provided by theorem 18 is best possible, after applying the ceiling function in the integers. Make a clear theorem statement about what this means and about what exactly you are proving.

4 Mathematical Induction

The principle of mathematical induction lies at the core of number theory and mathematics generally, serving as a powerful unifying central principle in axiomatic developments of the subject, the main engine used to prove essentially all of the fundamental facts of arithmetic and more. So let us explore this core principle a little more fully. We have already begun to use it informally in the previous chapters.

4.1 The least-number principle

The core idea of induction is expressed by the following axiomatic principle:

Principle 20 (Least-number principle). *If there is a natural number with a certain property, then there is a smallest number with that property.*

Equivalently, every nonempty set A of natural numbers has a least element. This principle asserts that the natural numbers are *well-ordered*, for a well-order is one for which every nonempty set has a least member.

The reader may have noticed that we made use of this principle in several previous arguments. We used it, for example, in chapter 1 in the revised proof that $\sqrt{2}$ is irrational, when we chose the numerator to be as small as possible, and we also used it in the proof that every rational number can be placed into lowest terms. We used the least-number principle again in the geometric proof that $\sqrt{2}$ is irrational, when we took the integer-sided squares to be as small as possible for which one had double the area of another, and we used it when proving the Euclidean algorithm and when proving the fundamental theorem of arithmetic. So although this chapter is titled, "Mathematical Induction," in fact we have already been making inductive arguments in the earlier chapters without saying so.

One should not think of mathematical induction as an obscure proof method used only to prove certain curious recursive sums. Rather, mathematical induction is a fundamental principle, a bedrock idea upon which all the basic facts of number theory rest. In the axiomatic developments of number theory from first principles, such as in the Peano or

Dedekind axiomatizations, for example, the principle of mathematical induction is used in almost every fundamental proof.

4.2 Common induction

In mathematical practice, elementary applications of induction often make use of certain idiosyncratic formulations of the least-number principle. Perhaps the most common formulation of the induction principle in elementary arguments is the following:

Theorem 21 (Common induction principle). *Suppose that A is a set of natural numbers for which* $0 \in A$, *and furthermore, whenever* $n \in A$, *then also* $n + 1 \in A$. *Then every natural number is in A.*

Another way to express the principle is that if 0 has a certain property, and the property is necessarily propagated from each natural number n to the next number $n + 1$, then indeed every natural number has the property.

The notation $x \in A$ appearing in the principle here means that x is an element of the set A. The claim that $0 \in A$ is commonly referred to as the *anchor* of the induction, and the implication $n \in A \implies n + 1 \in A$ is called the *induction step*. Because of the anchor, we know 0 is in the set. By applying to the induction step to that case, we see that 1 also is in the set. And by applying the induction step to *that* case, we see that 2 is in the set, and so on. Each number in the set causes the next number to be in the set. Like a train of dominoes, every number falls in turn into the set.

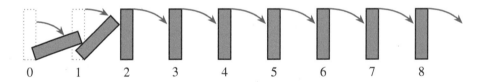

The first domino is pushed, and each falling domino causes the next to fall; in the end, of course, every domino falls. This image might help us see the fundamental correctness of the common induction principle; we shall nevertheless later prove the principle, using the least-number principle, and this is why I have stated the common induction principle as a theorem rather than as an axiom.

Alternative formulations of the common induction principle allow you to anchor the induction at a number other than 0. For example, if A is a set of integers for which $1 \in A$, and furthermore, whenever $n \in A$ then also $n + 1 \in A$, then every positive integer $n \geq 1$ is in A. We could also anchor the induction at 5, or at -3, or whatever, and conclude that all numbers from this anchor and above $n \geq 5$ are in the set. You will be asked to prove this in exercise 4.16. The principle of mathematical induction is robust and takes many useful forms.

4.3 Several proofs using induction

Before proving the common induction principle, let us first get a little practice using it. Consider the *triangular* numbers, which are the numbers of the form $1 + 2 + 3 + \cdots + n$. These numbers are called triangular because this many objects can be arranged in the form of a triangle, as pictured here, where one adds one more layer each time as n increases.

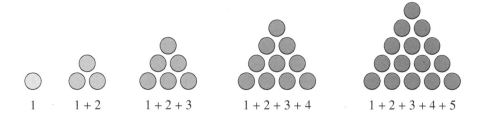

| 1 | $1 + 2$ | $1 + 2 + 3$ | $1 + 2 + 3 + 4$ | $1 + 2 + 3 + 4 + 5$ |

Theorem 22. *The sum of the first n positive integers is $n(n + 1)/2$, for any positive integer n. In other words,*

$$1 + 2 + 3 + \cdots + n = n(n + 1)/2.$$

Thus, the nth triangular number is $n(n + 1)/2$.

Proof. We prove the theorem by common induction. The statement is true at the anchor $n = 1$, since $1 = 1(1 + 1)/2$, so the number 1 is in the set of numbers for which the statement is true. Consider now the induction step. We assume that the statement is true for a number n and consider $n + 1$. Since the statement was true for the number n, we know that $1 + 2 + \cdots + n = n(n + 1)/2$. If we now add $n + 1$ to both sides of this equation and apply some elementary algebra, we see that

$$1 + 2 + \cdots + n + (n + 1) = n(n + 1)/2 + (n + 1)$$
$$= \frac{n^2 + n + 2n + 2}{2}$$
$$= (n^2 + 3n + 2)/2$$
$$= (n + 1)(n + 2)/2.$$

This is exactly what it means for the statement to be true in the case of $n + 1$, so we have proved that if the statement is true for n, then it is also true for $n+1$. Therefore, we conclude by the principle of induction that the statement is true for all positive integers. $\square$

We could have anchored the induction at $n = 0$ rather than $n = 1$, because the statement is also true for $n = 0$, as the sum of the first zero many positive integers is zero, which is equal to $0 \cdot 1/2$. In light of this, instead of stating the theorem as "for every positive integer n," we could have stated the theorem as "for any natural number n."

Theorem 23. *The sum of the first n odd numbers is n^2, for any natural number n. In other words,*

$$1 + 3 + 5 + \cdots + (2n - 1) = n^2.$$

Proof. Let us prove the theorem by common induction. The statement is easy to see when $n = 1$, since $1 = 1^2$. But again, we may actually take $n = 0$ as the base case in the original statement, since the sum of the first zero many odd numbers is 0, since there are not any, and $0^2 = 0$ as well, so it is true for $n = 0$. (Perhaps the case $n = 0$ is a little confusing in the case of the displayed equation, however, since perhaps it is not clear what the left side is supposed to indicate when $n = 0$. To my way of thinking, though, the only sensible interpretation of the left-side expression when $n = 0$ is that there are no terms in it at all, giving a sum of 0, which is indeed precisely what we want.) For the induction step, assume that the statement is true for a number n, and consider $n + 1$. Since the statement is true for n, we know that $1 + 3 + 5 + \cdots + (2n - 1) = n^2$. If we add $2(n + 1) - 1$ to both sides, which is the same as $2n + 1$, we get $1 + 3 + 5 + \cdots + (2n + 1) = n^2 + 2n + 1$, and this further simplifies to $(n + 1)^2$. Thus, the statement is true for $n + 1$. So we have proved that the statement is true at the anchor, and the truth propagates from each number n to $n + 1$. So by the principle of induction, we conclude that the statement is true for every positive integer. □

We shall have an alternative proof of theorem 23 in chapter 6.

Theorem 24. *The sum of the powers of 2 up to 2^n is 1 less than the next power of 2. In other words,*

$$1 + 2 + 4 + \cdots + 2^n = 2^{n+1} - 1.$$

Proof. We prove the theorem by common induction. The statement is true for $n = 0$, since $1 = 2^1 - 1$. Assume that the statement is true for a number n, and consider $n + 1$. Since the statement was true for n, we have $1 + 2 + 4 + \cdots + 2^n = 2^{n+1} - 1$. By adding 2^{n+1} to both sides, we get $1 + 2 + 4 + \cdots + 2^{n+1} = 2^{n+1} - 1 + 2^{n+1}$, which is equal to $2 \cdot 2^{n+1} - 1 = 2^{n+2} - 1$, which is what is desired for the case of $n + 1$. So the statement is true for all n by induction. □

The *Fibonacci sequence* is the sequence

$$0 \quad 1 \quad 1 \quad 2 \quad 3 \quad 5 \quad 8 \quad 13 \quad 21 \quad 34 \quad \cdots \quad .$$

Can you see the pattern? The sequence starts out with zero and one and then each next number is the sum of the previous two. We can therefore describe the sequence by the following recursive rules:

$$f_0 = 0$$
$$f_1 = 1$$
$$f_{n+2} = f_n + f_{n+1}$$

Theorem 25. *The Fibonacci numbers obey the following identity:*

$$(f_0)^2 + (f_1)^2 + \cdots + (f_n)^2 = f_n f_{n+1}.$$

Proof. We prove the theorem by common induction. The statement is true for $n = 0$, since $(f_0)^2 = 0 = 0 \cdot 1 = f_0 f_1$. Assume that the statement is true for a number n, and consider $n + 1$. Since the statement is true for n, we have $(f_0)^2 + \cdots + (f_n)^2 = f_n f_{n+1}$. Let us add $(f_{n+1})^2$ to both sides, concluding that $(f_0)^2 + \cdots + (f_n)^2 + (f_{n+1})^2 = f_n f_{n+1} + (f_{n+1})^2$, which is equal to $(f_n + f_{n+1}) f_{n+1}$. By applying the recursive definition, this is equal to $f_{n+2} f_{n+1}$. So we have established that $(f_0)^2 + \cdots + (f_{n+1})^2 = f_{n+1} f_{n+2}$, and so the statement is true for $n + 1$. Thus, by induction, it is true for all natural numbers n. $\square$

An alternative visual proof of this identity will be given in chapter 6.

Theorem 26. *The number $6^n - 1$ is a multiple of 5 for every natural number n.*

Proof. We prove the theorem by common induction. The statement is true when $n = 0$. Suppose by induction that the statement is true for a number n, and consider $n + 1$. Since $6^n - 1$ is a multiple of 5, we have $6^n - 1 = 5k$ for some integer k. In other words, $6^n = 5k + 1$. Multiplying both sides by 6, we see that $6^{n+1} = 6(5k + 1) = 30k + 6$, and consequently, $6^{n+1} - 1 = 30k + 5 = 5(6k + 1)$, which is a multiple of 5. So the statement is true for $n + 1$, and we have thereby proved by induction that the statement holds for all natural numbers. $\square$

The previous theorem can also be easily proved with modular arithmetic, for those who are familiar with it, because 6 modulo 5 is 1, and so the expression $6^n - 1$ is equivalent to $1^n - 1 \bmod 5$, which is $0 \bmod 5$. Thus, $6^n - 1$ is a multiple of 5.

Theorem 27. *The number $n^3 - n$ is a multiple of 6, for every natural number n.*

Proof. We prove the theorem by common induction. The statement is true when $n = 0$, since $0^3 - 0 = 0$, and this is a multiple of 6. Suppose by induction that the statement is true for a number n. So $n^3 - n$ is a multiple of 6. Observe that

$$(n + 1)^3 - (n + 1) = n^3 + 3n^2 + 3n + 1 - (n + 1)$$
$$= n^3 - n + 3n^2 + 3n$$
$$= (n^3 - n) + 3(n^2 + n)$$
$$= (n^3 - n) + 3n(n + 1).$$

The number $n(n + 1)$ is always even, since either n or $n + 1$ is even (as with the arguments of chapter 2), and so $3n(n + 1)$ is a multiple of 6. So the next number $(n + 1)^3 - (n + 1)$ arises from the previous number $n^3 - n$ by adding a multiple of 6. So if the previous number was a multiple of 6, then so will be the next number. So the statement is true for $n + 1$, and we have thereby proved by induction that the statement is true for all natural numbers. $\square$

This statement was also proved without induction in exercise 2.6.

Theorem 28. *The inequality $n^2 < 3^n$ holds for every natural number.*

Proof. We prove the theorem by common induction. First, the statement is easy to verify in the cases $n = 0$, $n = 1$, and $n = 2$, as the reader may easily check. Assume by induction that the statement is true for a number n, and consider $n + 1$. Since we have verified the claim up to 2 already, we may also assume that $2 \leq n$. Since the statement is true for this number n, we know that $n^2 < 3^n$. Using the fact that $2 \leq n$, we may argue with the following chain of inequalities:

$$
\begin{aligned}
(n + 1)^2 &= n^2 + 2n + 1 \\
&\leq n^2 + n^2 + n^2 \qquad &&\text{(using } 2 \leq n) \\
&= 3 \cdot n^2 \\
&< 3 \cdot 3^n \qquad &&\text{(by the induction hypothesis)} \\
&= 3^{n+1}
\end{aligned}
$$

And so the statement is true for $n + 1$. Thus, by mathematical induction, it holds for all natural numbers. □

I should like to call attention to the fact that in the proof of theorem 28 we made at first what might appear to be three different anchor cases, considering not only $n = 0$ but also $n = 1$ and $n = 2$. Why did we do this? Could we have skipped that part? Although with common induction one generally needs only one anchor, in this particular argument we actually needed those extra cases, because the argument we gave in the induction-step part of the proof began to work only once $2 \leq n$. One way to think about it is that we just established the cases $n = 0$ and $n = 1$ separately, but the real anchor was $n = 2$, since the induction step of the argument works only for $2 \leq n$.

4.4   Proving the induction principle

So we have stated the common induction principle, and we have used it in a few arguments. But can we prove the principle itself? Yes, in fact it is easy to prove the common induction principle from the least-number principle.

Proof of theorem 21. We prove the common induction principle using the least-number principle. Assume that A is an *inductive* set of natural numbers, which means that $0 \in A$ and whenever $n \in A$, then also $n+1 \in A$. We want to prove that every number is in A. If not, then there must be some numbers not in A, and by the least-number principle there must therefore be a smallest number that is not in A. Let us call this number m. The assumption on A ensures that $m \neq 0$. Consequently, $m \geq 1$, and so $m - 1$ is a natural number. Since $m - 1$ is smaller than m and m is the smallest number not in A, it must be that $m - 1$ is in A.

Therefore, by the induction assumption on A, it follows that $(m - 1) + 1$ is also in A. But this is m itself, and so m is in A after all, contradicting our earlier assumption that it was not. So A must contain every natural number. □

4.5 Strong induction

Although the common induction principle is indeed quite common, there are sometimes situations where one wants to prove something by induction, but the induction implication goes naturally not from n to $n + 1$ but, rather, from some other smaller number to $n + 1$, or from several smaller numbers. In these cases, the following stronger formulation of the induction principle is useful.

Theorem 29 (Strong induction principle). *Assume that A is a set of natural numbers with the property that, for every natural number n, if every number smaller than n is in A, then n itself is in A. Then every natural number is in A.*

One thing to notice about the strong induction formulation is that it has no anchor case! How can this induction principle be correct? But it *is* correct, I assure you, and the reason has to do with what is called *vacuous* truth. Specifically, in the case $n = 0$, there are no natural numbers smaller than n, and so it is vacuously true that all such numbers are in A, since this assertion can have no counterexample. So the strong induction assumption vacuously implies that $0 \in A$, and the induction engine is thereby started.

Proof of theorem 29. This argument is similar to the proof of the common induction principle from the least-number principle. Namely, assume that A satisfies the strong induction property mentioned in the strong induction principle. If not every natural number is in A, then there would be some least natural number n that is not in A. But in this case, every number smaller than n would be in A and so n itself would have to be in A after all, by the strong induction assumption on A. This contradicts our assumption on n, and so there cannot be any natural numbers not in A. In other words, every natural number must be in A, as desired. □

Several of our informal uses of mathematical induction that we gave in chapter 3 had essentially used strong induction. For example, to prove that every positive integer has a prime factorization (theorem 10), you suppose that every number smaller than n has such a factorization and then observe that either n is prime, in which case it is its own factorization, or else $n = ab$ for smaller numbers $a, b < n$, which by the induction assumption have prime factorizations, which can be put together to make a prime factorization of n. In this argument, you see, we did not move from n to $n + 1$ but, rather, reduced the problem on n to instances of the problem for smaller numbers a and b. Several of the other results in chapter 3 can similarly be set out as proofs by strong induction.

Let us now give a few more applications of the principle of induction. Consider the case of binary representation of numbers. Perhaps you know that the number 6 is represented in binary as 110, since $6 = 4 + 2$, and the number 19 is represented as 10011, since $19 = 16 + 2 + 1$. Perhaps you know that every natural number has a binary representation. But how could we prove this? Furthermore, how do we know that different binary representations always represent different numbers?

Theorem 30. *Every natural number can be expressed as the sum of a unique set of distinct powers of 2.*

Proof. Assume by strong induction that the statement is true for all numbers smaller than n. Since 0 is the empty sum, we may assume that $n \geq 1$. Let 2^k be the largest power of 2 that fits inside n, so that $2^k \leq n$. Since $n - 2^k$ is smaller than n, it has a unique representation as a sum of distinct powers of 2. Note that 2^k cannot appear in the sum for $n - 2^k$, since otherwise n would be at least $2^k + 2^k = 2^{k+1}$, but we had chosen 2^k to be the largest power of 2 that fits inside n. Thus, by adding 2^k to the sum for $n - 2^k$, we obtain n as a sum of distinct powers of 2, establishing the existence claim of the theorem. For uniqueness, note that 2^k must occur in any representation of n as a sum of distinct powers of 2, since no larger powers ever occur by the choice of k, and if we use only smaller powers, then even if we use all of them the sum will be at most $1 + 2 + 4 + \cdots + 2^{k-1}$, which by theorem 24 is less than 2^k and so therefore cannot sum to n. So 2^k is used in every representation of n as a sum of powers of 2, and the uniqueness claim therefore follows from the uniqueness of the representation of $n - 2^k$.   □

Corollary 31. *Every natural number has a unique binary representation.*

Proof. Every natural number has a unique binary representation, since the binary representation is simply a convenient compact notation for indicating which powers of 2 one is including in the sum, and by theorem 30 this collection is unique.   □

For example, the binary number 101101 is a compact notation for the sum

$$2^5 + 2^3 + 2^2 + 2^0,$$

which is 45 in decimal notation.

4.6   Buckets of Fish via nested induction

Let us consider now a somewhat more fanciful instance of induction, which also illustrates the idea of *nested* induction. Consider the game I call Buckets of Fish, played with two players on the beach. The game begins with finitely many buckets arranged in a row, each bucket containing some finite number of fish, with a large supply of additional fish available nearby, as many as needed, fresh off the boats. Each player, in turn, takes a fish from one bucket and adds to each of the buckets to the left, if any, as many fish from the extra supply as he or she likes. If we label the buckets 1, 2, 3, and so on, from left to right, then it would be a legal move for a player to take one fish out of bucket 4 and add ten fish (from the extra supply) to bucket 1, no fish to bucket 2, and ninety-four fish (from the extra supply) to bucket 3. The winner is whoever takes the very last fish from the buckets, leaving them empty.

Since huge numbers of fish can often be added to the buckets during play, but only one fish removed, a skeptical reader could reasonably wonder whether we'll really always even get a winner. By adding fish suitably, can the players prolong the game indefinitely? Or does every play of the game Buckets of Fish necessarily come to an end?

The answer is that, indeed, every play of the game must eventually come to an end. Regardless of how the players might conspire to add fish to the buckets during play, even with an endless supply of fish from the boats, nevertheless they will eventually run out of fish in the buckets and one of the players will take the last fish. I shall give several proofs.

Theorem 32. *Every play of the game Buckets of Fish ends in finitely many moves. All the fish in the buckets, including all the new fish that may have been added during play, will eventually run out by some finite stage during play.*

Before proving this theorem, let me first prove a weaker claim, namely, that either player may force the game to end in finitely many moves. The way to do this is simply to take fish always only out of the rightmost bucket. Since that bucket is never replenished, it follows that eventually it will become empty, and then there will be in effect strictly fewer buckets. By now taking fish only from the new rightmost bucket, we can eventually ensure that that bucket also becomes empty, and so on. In this way, by induction, we can ensure that the number of buckets with fish eventually reduces to just one bucket, which is eventually emptied. Thus, either player can play so as to ensure that the game ends in finitely many moves. But this is a weaker claim than made in the theorem, which states that actually all ways of playing the game will end in finitely many moves.

First proof. We prove the claim by (nested) induction on the number of buckets. If there is only one bucket, then there are no buckets to the left of it, and consequently there is no possibility in this case to add fish to the game; so if the one bucket contains k fish, then the claim clearly ends in k moves.

Assume by induction that all plays using at most n buckets end in finitely many moves, and suppose that we have a game situation with $n + 1$ buckets, with k fish in bucket $n + 1$. We now prove by induction on k that all such games terminate. This argument is therefore an instance of *nested* induction, because we are currently inside our proof by induction on n, in the induction step of that proof, and in order to complete it, we are undertaking a separate full induction on k. If $k = 0$, then there are no fish in bucket $n + 1$, and so the game amounts really to a game with only n buckets, which terminates in finitely many steps by our induction hypothesis on n.

So, let us assume that all plays with k fish in bucket $n + 1$ terminate in finitely many moves. Consider a situation where there are $k + 1$ many fish in that bucket. I claim that eventually one of those fish must be taken, since otherwise all the moves will be only on the first n buckets, and all plays on only n buckets terminate in finitely many moves. So at some point, one of the players will take a fish from bucket $n + 1$, possibly adding additional fish to the earlier buckets. But this produces a situation with only k fish in bucket $n + 1$, which by our induction assumption on k we know will terminate in finitely many steps.

So we have proved that no matter how many fish are in bucket $n + 1$, the game will end in finitely many moves, and so the original claim is true for $n + 1$ buckets. Thus, the theorem is true for any finite number of buckets.   □

Second proof. Let me now give another proof. We want to prove that there is no infinitely long play of the game Buckets of Fish. Suppose toward contradiction that there is a way for the players to conspire to produce an infinite play, starting from some configuration of some finite number n of buckets, each with finitely many fish in them.

Fix the particular infinitely long play. Let m be the rightmost bucket from which infinitely often a fish was taken during that infinite course of play. It follows, for example, that $m < n$, since bucket n can be used only finitely often, as it never gets replenished. Since bucket m starts with only finitely many fish in it, and each time it is replenished, it is replenished with only finitely many fish, it follows that in order to have been used infinitely many times, it must also have been replenished infinitely often. But each time it was replenished, it was because there was some bucket farther to the right that had been used. Since there are only finitely many buckets to the right of bucket m, it follows that one of them must have been used infinitely often. This contradicts the choice of m as the rightmost bucket that was used infinitely often.   □

Let me mention briefly a third proof of theorem 32, making use of the concept of transfinite ordinals, with which some readers might be familiar. (If you are not familiar with ordinals yet, please do not worry about it; you might learn about them in an introductory set theory class at your university.) Specifically, we associate with each Buckets-of-Fish position a certain ordinal. With the position

$$7 \quad 2 \quad 5 \quad 24,$$

for example, we associate the ordinal

$$\omega^3 \cdot 24 + \omega^2 \cdot 5 + \omega \cdot 2 + 7.$$

In general, the number of fish in each bucket of a position becomes the coefficient of the corresponding power of ω, using higher powers for the buckets farther to the right. The key observation to make is that these associated ordinals strictly descend for every move of the game, since one reduces a higher-power coefficient and increases only lower-power coefficients. Since there is no infinite descending sequence of ordinals, it follows that there is no infinite play in the game Buckets of Fish. This idea also shows that the ordinal game values of positions in this game are bounded above by ω^ω, and every ordinal less than ω^ω is realized by some position.

So now we know that the game will always end. But how shall we play? What is the winning strategy? Say you are faced with buckets having fish in the following amounts:

$$4 \quad 5 \quad 2 \quad 0 \quad 7 \quad 4$$

What is your winning move? Please give it some thought; we shall return to the topic in chapter 7, where we will provide the winning strategy in theorem 51.

4.7 Every number is interesting

I should like to conclude this chapter with an informal "proof" that every natural number is interesting. For example, 0 is interesting, since it is the additive identity, which is an interesting property for a number to have. Similarly, the number 1 is the multiplicative identity, which also is quite interesting. The number 2 is the first prime number, 3 is the first odd prime number, 4 is the first composite number, and so on.

I claim that *every* number is interesting. To prove this, observe that if there were any uninteresting numbers, then by the least-number principle, there would have to be a number n that was the smallest uninteresting number. But that is an extremely interesting property for a number to have! This therefore contradicts the assumption that n was uninteresting, and so there must not be any uninteresting numbers. So we may rejoice: every number is interesting! The reader will criticize this argument in exercise 4.18.

Mathematical Habits

Understand induction deeply. Grasp the idea that mathematical induction is about the impossibility of minimal counterexamples. Learn how this manifests in the various forms of induction—common induction, strong induction, the least-number principle—and understand deeply why these are valid.

Use induction flexibly. Do not insist upon a rigid format for proofs by induction but, rather, adopt a flexible approach, choosing whichever valid form of induction fits your argument best.

Distinguish sharply between example and proof. Key examples can be extremely valuable and insightful, but examples do not prove a universal statement. Examples can offer evidence in favor of a mathematical conjecture, but proving a universal claim will require a general argument. Examples can prove an existential statement, however, and they can refute a universal statement, in which case they are called counterexamples.

Exercises

4.1   Show by induction that $2^n < n!$ for all $n \geq 4$.

4.2   Show by induction that $\sum_{k=0}^{n} k \times k! = (n + 1)! - 1$.

4.3   Show by induction that $f_0 + \cdots + f_n = f_{n+2} - 1$ in the Fibonacci sequence.

4.4   Show by induction that $f_n < 2^n$ in the Fibonacci sequence.

4.5   Consider an alternative Fibonacci sequence, starting with $0, 2$. Can you prove the analogue of theorem 25? Generalize the result as far as you can.

4.6   In analogy with the triangular numbers, define the *pentagonal* numbers and the *hexagonal* numbers. Can you find and prove a formula for the nth pentagonal number and the nth hexagonal number?

4.7   Show by induction that a finite set with n elements has exactly 2^n many subsets.

4.8   Prove by induction that $1 + 4 + 4^2 + \cdots + 4^{n-1} = (4^n - 1)/3$ for every positive integer n.

4.9   Prove that $\frac{1}{1\cdot2} + \frac{1}{2\cdot3} + \frac{1}{3\cdot4} + \cdots + \frac{1}{n(n+1)} = \frac{n}{n+1}$.

4.10   Prove that $1 \cdot 2 + 2 \cdot 3 + 3 \cdot 4 + \cdots + n \cdot (n + 1) = \frac{n(n+1)(n+2)}{3}$.

4.11   Show that every natural number has a unique base 3 representation.

4.12   Prove that if you divide the plane into regions using finitely many straight lines, then the regions can be colored with two colors in such a way that adjacent regions have opposite colors. [Hint: Use induction on the number of lines.]

4.13   True or false: every flying polka-dotted elephant in this room is smoking a cigar.

4.14   Criticize the following "proof." Claim. All horses are the same color. Specifically, every finite set of horses is monochromatic. Proof. We argue by induction. The statement is clearly true for sets of size 1. Assume by induction that all sets of n horses are monochromatic, and consider a set of size $n + 1$. The first n horses are all the same color. The last n horses are all the same color. Because of the overlap, this means that all $n + 1$ horses are the same color. So by induction, all finite sets of horses are all the same color, and so all horses are the same color. $\square$

4.15 Criticize the following "proof." Claim. $\sum_{i=1}^{\infty} \frac{1}{i} < \infty$. Proof. Let $S(n) = \sum_{i=1}^{n} \frac{1}{i}$, and consider the statement asserting $S(n) < \infty$. This statement is true for $n = 1$, since $\frac{1}{1} = 1 < \infty$. And if it is true for n, then it is true for $n + 1$, since the next sum $S(n+1) = \sum_{i=1}^{n+1} \frac{1}{i}$ is equal to the previous sum $\sum_{i=1}^{n} \frac{1}{i}$ plus the next term $\frac{1}{n+1}$, or in other words, $S(n+1) = S(n) + \frac{1}{n+1}$, and this is the sum of two finite numbers. So $S(1)$ is true, and $S(n)$ implies $S(n+1)$. So we have therefore proved $\sum_{i=1}^{\infty} \frac{1}{i} < \infty$ by induction. $\square$

4.16 Prove the following common induction principle variation: Assume that A is a set of integers with some $k \in A$, and furthermore, that whenever $n \in A$, then also $n + 1 \in A$. Then A contains all integers above k.

4.17 Prove that the least-number principle, the common induction principle, and the strong induction principle are all equivalent. Assuming any one of them, you can prove the others.

4.18 Criticize the argument that every number is interesting, given at the end of the chapter.

Credits

Exercise 4.15 was proposed by an anonymous user on the mathematics question-and-answer website math.stackexchange.net, User3203476 (2016). The second proof of theorem 32 arose in a discussion I had with Miha Habič.

5 Discrete Mathematics

5.1 More pointed at than pointing

Gather yourself and a few friends into a circle and point at each other in some arrangement of pointing. Let each person point at one or more of the others, or at themselves, or at nobody, as they like. Use both hands, or different fingers, or your feet if you want to point at several people, and let us say that it is allowed to point more than once at a given person, or at several people, or at yourself—go to town! Perhaps some people are pointing quite a lot and others are pointing much less, and similarly with being pointed at. Now, I have a

question about whether you might be able to achieve a certain feature in your pattern of pointing at each other.

Question 33. Can you arrange it so that every person is altogether more often pointed at than pointing?

In other words, could we all be pointed at strictly more times than we point at others? Ponder the problem on your own, before reading further.

Interlude...

Theorem 34. *The answer is no, it is not possible to have a nonempty finite set of people pointing at each other in such a way that every person is more often pointed at than pointing.*

Let us give several different proofs.

First proof. Suppose that we have a finite arrangement of people pointing at each other or themselves. For each person, let us say their *pointed-at score* is the number of times someone is pointing at them, and their *pointing score* is the number of times they are pointing at someone, including all instances of multiple pointing and self-pointing in both of these scores. Let A be the sum total of all the pointed-at scores, and let P be the sum total of all the pointing scores. I claim that $P = A$. The reason is that every instance of someone pointing is also simultaneously an instance of someone being pointed at, simply viewed from the other person's perspective, at the other end of the finger. Every instance of pointing adds exactly one to P and also exactly one to A. If every person were more often pointed at than pointing, however, then it would follow that $P < A$, since P would be the sum of a finite sequence of numbers, each of which is smaller than the corresponding summands giving rise to A. Since $P = A$, this cannot happen.   □

Next, we prove the theorem inductively.

Second proof. We prove the theorem by induction on the number of people. That is, no set of n people can form a counterexample. This statement is true for $n = 1$ person, since the person can point only at herself, and if she does so k times, then she will be both pointing and pointed at k times equally. Suppose now that the statement is true for all groups of size n, and consider a group of $n + 1$ people. Suppose that we have an arrangement of the $n + 1$ people for which everyone is more often pointed at than pointing. Let us call one of those people "Horatio." In particular, Horatio is more often pointed at than pointing. Thus, we may simply remove Horatio from the group of people and direct some of the people who were pointing at him to point instead at those to whom Horatio had pointed. Since Horatio was more often pointed at than pointing, there are enough people who had been pointing at Horatio to cover his pointing commitment. After this rearrangement of the pointing, anyone left still pointing at where Horatio had been may simply lower his or her finger. In this way, we arrive at a new configuration, with one fewer person and hence of size n, but that still satisfies that everyone left is more often pointed at than pointing. This contradicts the induction assumption that there is no such group of size n, and so we have completed the induction step. So there can be no such group of people of any finite size.   □

In the third proof, let us adopt an anthropomorphizing perspective, which enables us more easily to see the truth of a certain mathematical feature.

Third proof. Suppose we are part of a finite group of people pointing at each other, and everyone is more often pointed at than pointing. Let us instruct everyone to pay one dollar each to the people to whom they point, for each instance of pointing; and let us assume that we all have enough cash on hand to do this. The curious thing to notice is that, after the payments, because everyone is more often pointed at than pointing, it follows that every person will take in more money than they paid out. We made money! And we could do it again and make more money, and again and again, as many times as we desire. We could make millions of dollars simply by exchanging it like this. Since this is clearly impossible, as the total dollar holdings of the group does not change as money is exchanged within it, there can be no such pointing arrangement. □

I find the third proof very clear, though I recognize that it is essentially similar to the first proof, if one simply thinks of the pointing-at and pointing scores as measured in dollars. Perhaps the reason it is so clear is that it replaces the abstract quantity-preserving argument of the first proof with something much easier to grasp, namely, the fundamental fact that we cannot get more money as a group simply by exchanging money within our group. Such anthropomorphizing arguments or metaphors can often be surprisingly effective in simplifying a mathematical idea. We leverage our innate human experience in order to understand more easily what would otherwise be a complex mathematical matter. Our human experience with the difficulty of getting money makes the final conclusion of the third argument obvious.

5.2 Chocolate bar problem

Consider next the chocolate bar problem. Imagine a rectangular chocolate bar, the kind having a pattern of small squares. We shall break the chocolate along these lines, in such a way that in the end we have only those tiny squares as separate pieces.

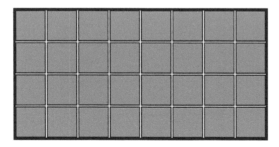

There are a variety of ways that we might do this. For example, for the bar pictured above, we could first make the three long breaks, making four 8×1 sticks, and then would break off one square at a time from those sticks. This would make $3 + 4 \cdot 7$ breaks altogether. Alternatively, we could first make all the short breaks, and then break off individual squares from the resulting 1×4 sticks, resulting in $7 + 8 \cdot 3$ breaks.

Question 35. What is the most efficient method of breaking the chocolate bar into squares, using the fewest total number of breaks?

Let us be honest in our counting of what a *break* means; we are not allowed to break two pieces off at once or break off an empty piece. To break a piece of chocolate means to take a single connected piece of chocolate and separate it into two nonempty pieces by cutting along one of the lines between the squares, following the line all the way across. How would you break the chocolate bar? Does it matter how you do it? In fact, it does not.

Theorem 36. *Breaking a chocolate bar into individual squares always takes exactly the same number of steps, regardless of the breaking protocol that is followed.*

Proof. Notice that each time we break a rectangle along an edge, we make two rectangles, each a bit smaller than the original. Each break increases the number of rectangles by exactly one. If the original piece of chocolate has n small squares, therefore, then after $n - 1$ breaks, regardless of how the breaks are performed (as long as each break creates exactly one new piece), there will be n pieces. And in this case, each piece must be a single small square. So all methods of breaking the chocolate bar use the same number of breaks, which is one less than the number of squares in the bar. □

5.3 Tiling problems

Consider next a tiling problem using small L-shaped tiles on a large board.

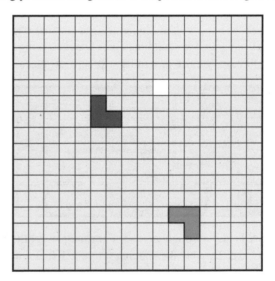

Theorem 37. *Every $2^n \times 2^n$ grid of unit squares, with one square removed, can be tiled by L-shaped tiles consisting of three unit squares.*

Proof. We prove this by induction on n. The claim is clear when $n = 0$, since in this case we have a 1×1 grid, with only one square, and when we remove it, we can tile what remains with no tiles at all. (The case $n = 1$, or a 2×2 grid, is similarly trivial, since when you remove one of the squares, what is left is perfectly covered by a single tile.)

Suppose now by induction that the statement is true for a number n, and consider a $2^{n+1} \times 2^{n+1}$ grid of unit squares, with one unit square removed. Since 2^{n+1} is even, we may divide the large square into four $2^n \times 2^n$ quadrants. The omitted square resides in one of them. Place a single tile near the center, oriented so that it covers one corner square from each of the other three quadrants, as indicated:

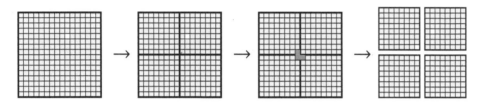

If we now think of the smaller quadrants separately, we have four smaller quadrants, each with one square removed. By the induction hypothesis, these can each be tiled with the L-shaped tiles. Together with our initial center tile, we thereby produce a tiling of the whole square. So all such grids with one square omitted can be tiled. ☐

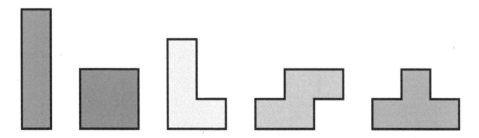

Consider next the collection of shapes, similar to those in the popular game Tetris, that can be formed from four unit squares. There are five such shapes, pictured above.

Question 38. Can you arrange these shapes to form a rectangle?

For example, can you tile a 4×5 rectangle with these shapes, using one tile of each type? Give it a try!

Interlude...

Did you try it? The answer is no, one cannot tile a rectangle with these shapes.

Theorem 39. *The collection of tiles, with one of each tile shape type above, cannot tile a rectangle.*

Proof. Suppose that we could tile a rectangle using exactly one of each type of those tile shapes. Since we have five shapes, each with area four, the total area is twenty, and so the possible size rectangles are 4×5, 2×10, or 1×20. One can see easily that the latter two rectangles are impossible, but our argument works generally, and so there is no need to argue separately for that case. Imagine that we have such a tiling of the rectangle. Let us place a chessboard pattern on the rectangle, and note that there are equal numbers of dark and light squares (but see also exercise 5.9).

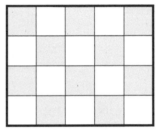

This chessboard pattern will induce a chessboard pattern on each of the pieces. Namely, if we could arrange the pieces to form the rectangle, then they would each inherit a light-square/dark-square pattern from the global chessboard pattern.

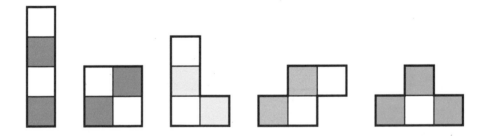

The thing to notice, however, is that each of the individual tiles would have exactly two dark squares and two light squares, with the exception of the final inverted T-shaped piece, which has three dark squares and one light square, or possibly three light squares and one dark square, depending on how it should happen to land on the chessboard pattern. This means that there cannot be a tiling of the rectangle, because on the one hand, the number of light and dark squares on the rectangle is equal, but when you think about the individual tiles, the problematic green tile will prevent the light and dark squares from balancing. So there can be no tiling. □

5.4 Escape!

Consider next the game Escape! We have three stones in the corner of an infinite quarter plane of squares. The rule of movement is that you can select any stone you like, and it will split into two stones, one moving to the square above and one moving to the square to the right of where it had been. The move is allowed only when both of those squares are empty, so that they may accept the new stones. The goal is to vacate the shaded L-shaped region at the origin.

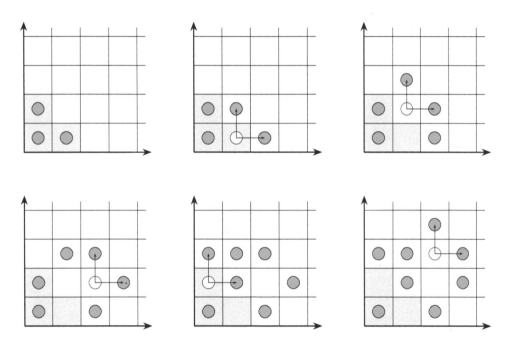

Can you vacate the shaded corner area? Please give it a try. One can make a good start, but then the outer stones begin to block one's path. You have to move these other stones out of the way to make room. Is it possible to get all three stones out of the corner? I have provided an online interactive version of the game, for trying out strategies and ideas, at https://scratch.mit.edu/projects/195391196, and see also my blog post at http://jdh.hamkins.org/escape.

Interlude...

Did you succeed? Perhaps one begins to think that it is impossible. But how could one ever prove such a thing?

Theorem 40. *It is impossible to make moves in the Escape! game so as to vacate the shaded corner region.*

Proof. Let us assign weights to the squares in the lattice according to the following pattern: We give the corner square weight 1/2, the next diagonal of squares 1/4 each, and then 1/8, and so on throughout the whole playing board. Every square gets a corresponding weight according to the indicated pattern.

$\frac{1}{32}$				
$\frac{1}{16}$	$\frac{1}{32}$			
$\frac{1}{8}$	$\frac{1}{16}$	$\frac{1}{32}$		
$\frac{1}{4}$	$\frac{1}{8}$	$\frac{1}{16}$	$\frac{1}{32}$	
$\frac{1}{2}$	$\frac{1}{4}$	$\frac{1}{8}$	$\frac{1}{16}$	$\frac{1}{32}$

The weights are specifically arranged so that making a move in the game preserves the total weight of the occupied squares. That is, the total weight of the occupied squares is invariant as one makes moves, because moving a stone with weight $1/2^k$ will create two stones of weight $1/2^{k+1}$, which adds up to the same. Since the original three stones have total weight $\frac{1}{2} + \frac{1}{4} + \frac{1}{4} = 1$, it follows that the total weight remains 1 after every move in the game. Meanwhile, let us consider the total weight of all the squares on the board. If you consider the bottom row only, the weights add to $\frac{1}{2} + \frac{1}{4} + \frac{1}{8} + \cdots$, which is the geometric series with sum 1. The next row has total weight $\frac{1}{4} + \frac{1}{8} + \frac{1}{16} + \cdots$, which adds to $1/2$. And the next adds to $1/4$ and so on. So the total weight of all the squares on the board is $1 + \frac{1}{2} + \frac{1}{4} + \cdots$, which is 2. This leaves a total weight of 1 for the unshaded squares. The subtle conclusion is that after any finite number of moves, only finitely many of those other squares are occupied, and so some of them remain empty. So after only finitely many moves, the total weight of the occupied squares off the original L-shape is strictly less than 1. Since the total weight of all the occupied squares is exactly 1, this means that the L-shape has not been vacated. So it is impossible to vacate the original L-shape in finitely many moves.   □

5.5 Representing integers as a sum

Let us investigate how many ways we can represent a given number as a sum.

Theorem 41. *Every positive integer n can be expressed as a sum of one or more positive integers in precisely 2^{n-1} many ways.*

For example, the number 4 can be expressed in the following eight ways:

$$1 + 1 + 1 + 1 \qquad\qquad 2 + 2$$
$$2 + 1 + 1 \qquad\qquad 3 + 1$$
$$1 + 2 + 1 \qquad\qquad 1 + 3$$
$$1 + 1 + 2 \qquad\qquad 4$$

For the purpose of this theorem, we allow the sum to have only one term, if desired, or many terms, and we pay attention to the order of the summands, but we do not use parentheses or pay attention to associativity.

Proof. Place n ones down in a row, like a picket fence:

$$1 \quad 1 \quad 1 \quad \cdots \quad 1 \quad 1$$

There are $n - 1$ spaces between the ones. For each such space, let us imagine placing either a $+$ sign in it or leaving it empty, in all the various possible ways to do this. Since there are $n-1$ spaces and two choices for each space, there are precisely 2^{n-1} many different ways to place these signs. Each such placement gives rise to a distinct way to represent n as a sum, if we interpret each contiguous block of ones as a single summand, so that

$$1 + 1 \quad 1 \quad 1 + 1 \quad 1,$$

for example, represents $1 + 3 + 2$. And conversely, every representation of n as a sum corresponds to such a marking. So there are 2^{n-1} ways to represent n as a sum. $\square$

A *fence-post error* is a common kind of mathematical error conflating fence posts with the spaces between them, such as in a list of numbers, an array, or an interval of time, and arises because there is one fewer space between the fence posts than there are fence posts. For example, if you arrive at a hotel on Sunday and leave the following Saturday, then you were present at the hotel on all seven days of the week, but you stayed only six nights. There are five integers from 8 to 12, even though these numbers differ by only four. If you were president of the club from the beginning of 2014 to the end of 2016, then you were president for three years: 2014, 2015, and 2016.

In exercise 5.11, the reader will give an alternative proof of theorem 41 by induction.

5.6   Permutations and combinations

A *permutation* of a list of objects is a rearrangement of those same objects into a different order. For example, there are six permutations of a list of three objects:

$$abc \quad acb \quad bac \quad bca \quad cab \quad cba$$

More generally, mathematicians define that a permutation of a set is a one-to-one correspondence of the set with itself. In the case of a finite set, the permutations in effect describe all the different ways that we might enumerate the elements of the set.

For any natural number n, we define the *factorial* number $n!$ by recursion:

$$0! = 1$$
$$(n + 1)! = (n + 1) \cdot n!.$$

In this way, one can see that $n!$ is simply the product of all the numbers from n down to 1, as in $5! = 5 \cdot 4 \cdot 3 \cdot 2 \cdot 1 = 120$.

Theorem 42. *For any integer $n \geq 0$, the number of permutations of n objects is the factorial $n!$.*

Proof. We prove the theorem by common induction. The statement is true when $n = 0$, since there is exactly one arrangement of zero objects, the empty arrangement. If the reader finds it confusing to consider permutations of the empty set, it may help to observe that the theorem is also true when $n = 1$, since there is exactly one way to permute one object also, and $1! = 1$.

Now, suppose by induction that the theorem is true for a number n, and consider $n + 1$. If have $n + 1$ objects $a_1, a_2, \ldots, a_n, a_{n+1}$ to be rearranged, let us first rearrange the first n objects $a_1, \ldots a_n$. By the induction hypothesis, there are $n!$ many ways to do that. Next, we place the final object a_{n+1} into the list. But where? There are precisely $n+1$ many slots into which it might be placed: before all them, between two of them, or after all of them. Every rearrangement of the $n + 1$ objects can be realized by first rearranging the first n objects and then placing the final object into one of the slots. So we will have $(n + 1) \cdot n!$ many rearrangements of $n + 1$ objects, and this is precisely $(n + 1)!$, as desired. So by induction, the theorem is true for all numbers n.   □

Why did we define $0! = 1$? Does that seem unnatural? Students sometimes suggest or expect that $0!$ should be 0. Would that be a good idea? As mathematicians, we are free to define our functions however we want. Which definition is best?

Notice that if we set $0!$ to be 0, then it would break the identity $(n+1)! = (n+1) \cdot n!$ in the case $n = 0$; we might have to start adding exceptions to our theorems on account of this. Indeed, the previous theorem itself provides a very good reason to define $0! = 1$, since as we observed, there is exactly one arrangement of the empty set, the empty arrangement. For this reason, among others, mathematicians have agreed that it is best to define $0! = 1$.

For integers $0 \leq k \leq n$, we define $\binom{n}{k}$, pronounced, "n choose k," as the number of ways to choose k items from a set of n items, disregarding the order.

Theorem 43. *The number of ways to choose k items from a set of n items, for $0 \leq k \leq n$, is given by the formula*

$$\binom{n}{k} = \frac{n!}{k!(n-k)!}.$$

We shall give several proofs.

First proof of theorem 43. Consider the following process: enumerate all n items, and then take just the first k in that enumeration. We will often get the same set of k elements, so there is considerable multiple counting in this process. But how much? The number of enumerations of n objects is precisely $n!$, by theorem 42. For a given enumeration of the objects, there are $k!$ many permutations of the first k elements of it, and $(n-k)!$ many permutations of the other elements. So there are $k!(n-k)!$ many ways to permute the elements in the enumeration, while having the same first k elements. So this is the factor that we have double counted by, and so the number of ways to choose k elements from n is precisely $n!/k!(n-k)!$, as desired. $\qquad\square$

The second proof will make use of the following lemma:

Lemma 44. $\binom{n+1}{k+1} = \binom{n}{k} + \binom{n}{k+1}$.

Proof. Consider a set A with $n+1$ objects, and we wish to count the number of ways of choosing $k+1$ objects from A. Fix a particular element $a \in A$. Now, some of the size-$(k+1)$ subsets of A have the element a, and some may not. There are exactly $\binom{n}{k}$ many size-$(k+1)$ subsets of A that have the element a, since once you choose a, then you must choose k additional elements from $A \setminus \{a\}$, a set of size n. Similarly, there are exactly $\binom{n}{k+1}$ many ways to choose $k+1$ elements from A without using the element a, since in this case one is really choosing from amongst the other n elements. So the total number of ways of choosing $k+1$ elements from A is the sum of these two, and we have proved the lemma. $\qquad\square$

Second proof of theorem 43. We prove the theorem by induction on n. If $n=0$, then also $k=0$, and it is easy to verify that there is exactly one way to choose nothing from nothing, so $\binom{0}{0} = 1$, which fits the formula $\frac{0!}{0! \cdot 0!}$, and so the anchor case is proved. Suppose now that the formula is correct for a number n, using any k, and consider the next number, $n+1$. Since again there is only one way to choose nothing from an n element set, we see that $\binom{n+1}{0} = 1$, which is equal to $\frac{n!}{0! n!}$, verifying the theorem in this case. Next, we consider $\binom{n+1}{k+1}$. By lemma 44, this is equal to $\binom{n}{k} + \binom{n}{k+1}$. By the induction hypothesis, we know that $\binom{n}{k} = n!/k!(n-k)!$ and $\binom{n}{k+1} = n!/(k+1)!(n-(k+1))!$.

Putting all of this together, we may now compute

$$\binom{n+1}{k+1} = \binom{n}{k} + \binom{n}{k+1}$$

$$= \frac{n!}{k!(n-k)!} + \frac{n!}{(k+1)!(n-(k+1))!}$$

$$= \frac{n!(k+1) + n!(n-k)}{(k+1)!(n-k)!}$$

$$= \frac{n!(k+1+n-k)}{(k+1)!(n-k)!}$$

$$= \frac{n!(n+1)}{(k+1)!(n-k)!}$$

$$= \frac{(n+1)!}{(k+1)!((n+1)-(k+1))!}.$$

This verifies the required instance of the theorem, and so we have proved it by induction for all natural numbers.   □

5.7 The pigeon-hole principle

Let us now turn to the following fundamental principle of combinatorics:

Theorem 45 (Pigeon-hole principle). *There is no one-to-one correspondence between a set of n elements and a set of k elements if k < n.*

Let me begin by saying that, although many people readily agree to the truth of the pigeon-hole principle, the fact of the matter is that we seem to have ample empirical evidence *against* it. Imagine two people are faced with a mountainous heap of pennies, and they both start counting them. Are you confident they will get exactly the same answer? If they get different answers, then is this not evidence against the pigeon-hole principle? The two ways of counting off the pennies would provide a one-to-one correspondence of the two numbers. In close elections, there are sometimes recounts of the votes, and it is rare that the count comes out exactly the same. Is this not evidence against the pigeon-hole principle?

Meanwhile, let us prove the theorem. The pigeon-hole principle is indeed correct, despite the ample empirical evidence against it. I suppose we should conclude that with very large numbers of things, people can make mistakes in counting. We shall investigate the concept of one-to-one correspondence further in chapter 11.

Proof of theorem 45. We argue by induction on *n*. The statement is vacuously true when *n* = 0, since there are no natural numbers *k* < 0. Suppose that the statement is true for a number *n*, and consider *n* + 1. Suppose that *A* is a set with *n* + 1 many elements, and that *j* : *A* → *B* is a one-to-one correspondence between *A* and a set *B* of size less than

$n + 1$, which is to say, of size at most n. Fix any particular element $a \in A$, so that the set $A' = A \setminus \{a\}$ has one fewer element, or n elements. The corresponding element $j(a)$ is in B, and we may let $B' = B \setminus \{j(a)\}$, which has size one less than B. Thus, B' has size less than n. Notice that the restriction of j to the set A' gives a correspondence $j \upharpoonright A' : A' \to B'$ of A' with B'. Furthermore, this restricted correspondence remains a one-to-one correspondence. Since A' has size n, this contradicts the induction hypothesis, and so there could have been no such correspondence j in the first place. $\square$

5.8 The zigzag theorem

Let us conclude with a fun observation that I call the zigzag theorem. Consider any rectangle, and draw a zigzag pattern in it, moving back and forth from the bottom edge to the top along straight lines, as many times as you like (as indicated below) but without crossing your own lines. This forms a number of triangles. What proportion of area is in the triangles below the zigzag pattern?

Theorem 46. *The triangles formed by any zigzag pattern in a rectangle carves out exactly half the area of the rectangle.*

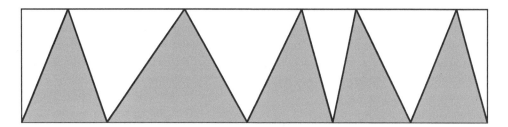

Proof. We assume at first in this proof that the zigzag pattern never doubles back, so that the zigzag path proceeds steadily to the right. In this case, all the triangles are acute, and the vertex of each triangle occurs above its base. Let us construct a vertical line on each vertex used in the zigzag pattern. This carves the rectangle into smaller rectangles, as illustrated below, and the main point is that each such rectangle is cut exactly in half diagonally by the corresponding zig or zag across the rectangle.

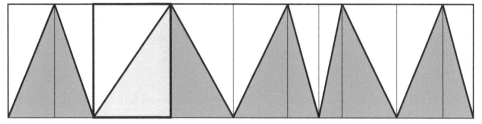

Thus, exactly half the area of the rectangle is above the zigzag and half below, as desired. This proves the theorem for the acute-triangle case. Let us now drop the no-doubling-back assumption and consider the general case, where we allow zigzag patterns that sometimes move backward (but not crossing themselves), like this:

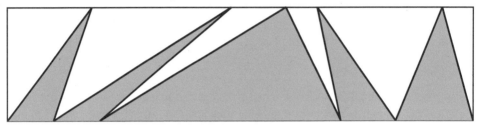

This kind of case breaks the previous proof, because the constructed rectangles would overlap. Nevertheless, we can mount an alternative argument. The point is that if the original rectangle has width w and height h, then each triangle has height h and thus area $\frac{1}{2}bh$, where b is the base of that triangle. Since the triangle bases add up to the total width w, it follows that the sum of the areas of the triangles is $\frac{1}{2}wh$, which is exactly half the area of the rectangle, as desired. $\qquad\square$

Alternative proof. Another way for us to argue in the general case is to observe that moving the top vertices of the triangles along the top edge of the rectangle is an area-preserving transformation because it preserves the base and height of each triangle. We can transform the skewed-triangle zigzags to the acute-triangle case, for example, simply by moving the top vertex of every triangle to be above its base. More directly, however, let us simply combine all the triangles into one by moving their top vertices all the way to the right like this:

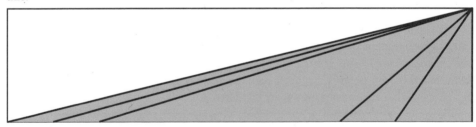

After this transformation, they form the region below the diagonal of the rectangle, which is exactly half the area. Since the transformation is area preserving, the original triangles must also have exactly half the area. $\qquad\square$

I find it interesting to notice that the theorem remains true even for zigzag patterns involving infinitely many zigs and zags. There is no need in the argument to assume that there were only finitely many triangles.

Mathematical Habits

Recognize when you have a proof and when you do not. This is an important, difficult step in one's mathematical development. Unfortunately, beginners are sometimes satisfied by an argument that experienced mathematicians will say is nonsense. Perhaps the attempted argument makes unwarranted assumptions, or misuses terms, or does not logically establish the full conclusion, or perhaps it makes any number of other mathematical errors, without the beginner realizing this. Therefore, be honest with yourself in your work; be skeptical about your argument; avoid talking nonsense. If you are not sure whether you have a proof, then you probably do not. Verify that your arguments logically establish their conclusions. Do not offer hand-waving arguments that avoid difficult but essential details. Never offer arguments that you do not understand.

Use metaphor. Express your mathematical issues metaphorically in terms of a familiar human experience, if doing so makes them easier to understand. Find evocative terminology that represents your mathematical quantities or relationships in familiar terms, if doing so supports the mathematical analysis.

Make conjectures. Use your mathematical insight, based on examples or suggestive reasoning, to guess the answer to a mathematical question or the mathematical fact that would explain a given mathematical phenomenon. Test your conjecture by checking whether it is consistent with known facts or examples. Try to prove your conjecture.

Exercises

5.1 Suppose that a finite group of people has some pattern of pointing at each other, with each person pointing at some or all or none of the others or themselves. Prove that if there is a person who is more often pointed at than pointing, then there is another person who is less often pointed at than pointing.

5.2 Suppose that you could control who follows whom on Twitter. Could you arrange it so that every person has more followers than people they follow? For example, some extremely famous people currently have many millions of followers, and one might hope to reassign most of those followers in such a way that everyone will be more followed than following.

5.3 Show that if there are infinitely many people, then it could be possible for every person to be more pointed at than pointing. Indeed, can you arrange infinitely many people, such that each person points at only one person but is pointed at by infinitely many people? How does this situation interact with the money-making third proof of theorem 34?

5.4   Prove that if the initial chocolate bar in theorem 36 is a rectangle, then one always has rectangular pieces after every stage of the breaking process.

5.5   Generalize the chocolate bar theorem (theorem 36) to nonrectangular chocolate bars.

5.6   Is the conclusion of theorem 37 true for $n \times n$ squares in general? Why or why not?

5.7   Explain how the proof of theorem 37 provides a construction method for producing the desired tiling. (Namely, once a given square is omitted, then perform the division into quadrants, place the one tile, and iterate with the smaller squares.) What tiling do you get this way for the 16×16 grid shown in the illustration?

5.8   Prove that the collection of shapes pictured before theorem 39 is indeed the collection of all shapes in the plane, freely allowing rotations and reflections, that can be formed from four unit squares by joining them at vertices along their edges.

5.9   Prove or refute the following statement: If you place a chessboard pattern on an $n \times m$ rectangular grid, there will be an equal number of dark and light squares. Generalize your answer as much as you can.

5.10   In the context of theorem 39, suppose that you have three pieces of each shape. Can you now tile a rectangle? Can you tile a 10×10 square with five tiles of each type?

5.11   Give an alternative proof of theorem 41 by induction. [Hint: Consider how to generate representations of $n + 1$ from representations of n.]

5.12   In the Escape! game, is it possible to fill as much as desired of the plane outside of the original L-shape? For example, can one make sure for any given n that all squares in the $n \times n$ square, except possibly the original L-shape, have a stone?

5.13   Consider a version of the Escape! game on a finite $n \times n$ board, modifying the rules so that stones that would have been placed outside this region are simply not placed. For which values of n can you vacate the yellow corner? For which values of n can you vacate the entire board?

5.14   Show that in the $n \times n$ version of Escape!, you cannot lose: every sequence of legal moves leads eventually to a completely vacated board. [Hint: Show first that a nonempty board always has a legal move; next, show by induction from the lower left that every square can be activated only finitely many times. So there is no infinite play, and therefore the board must become empty at a finite stage of play.]

5.15   (Challenge) Show that the number of steps to vacate the $n \times n$ board does not depend on the particular sequence of moves that are made. The game always ends with an empty board in exactly the same number of steps regardless of how the plays are made.

Credits

Theorem 37 is due to Solomon Golomb (1954), who also investigated similar tilings of many other shapes. Theorem 41 was observed in 1951 by Arthur B. Brown of Queens College and proved by William Moser of the University of Toronto. The pointing-at painting is *David and Charles Colyear* by Sir Godfrey Kneller, in the public domain via Wikimedia Commons. The Escape! game is due to Maxim Kontesevic and was discussed on the Numberphile video series at https://www.youtube.com/watch?v=lFQGSGsXbXE&t=11s.

6 Proofs without Words

Mathematics calls for polyglots, for we express our mathematical ideas in many languages, conveying them with algebraic equations, by inequalities, in formal languages, in natural language, or with figures. Indeed, many mathematical concepts are more readily expressed visually in a diagram or figure than with words or symbols, and mathematicians therefore draw upon such visual aids to clarify or even to express their mathematical ideas. My advice is to include insightful pictures or diagrams in a mathematical argument whenever possible.

In an extreme form of this, the genre of mathematical proof known as *proofs without words*, mathematicians attempt to communicate the essence of a mathematical proof using no words at all, conveying their idea entirely by means of an insightful graphic or diagram. These arguments can be very clever, with the figure alone often sufficing to communicate the proof idea. To figure out the intended mathematical argument can be an enjoyable mathematical puzzle, often leading to a satisfying *Aha!* moment.

6.1 A geometric sum

Let us begin with an easy example. Consider the figure here, which aims to prove the geometric series sum

$$\frac{1}{2} + \frac{1}{4} + \frac{1}{8} + \frac{1}{16} \cdots = 1.$$

Can you see the intended argument? As a "proof without words," the figure is meant to stand on its own as a proof of the identity. Try to figure out how the figure can be used to prove the identity.

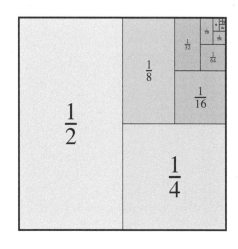

Interlude...

Aha! Did you discover the intended proof? The idea, of course, is that the whole unit square has area 1, and it is successively divided into smaller pieces: a rectangle of area $\frac{1}{2}$, a square of area $\frac{1}{4}$, and so on. Each rectangle is followed by a square of half the size, and each square is followed by a rectangle of half the size. So the sum of all the pieces is $\frac{1}{2} + \frac{1}{4} + \frac{1}{8} + \frac{1}{16} + \dots$, and since they exhaust the unit square in the limit, the sum is 1, as desired.

6.2   Binomial square

For another easy case, consider the following diagram, offered as a proof of the binomial square identity. Can you see how the diagram proves the identity?

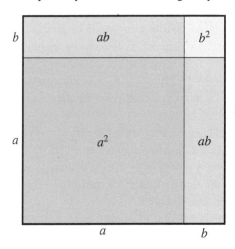

$$(a + b)^2 = a^2 + 2ab + b^2$$

Aha! The whole square, having sides $a + b$, has total area $(a + b)^2$; but this area can also be realized as the sum of four regions, two squares and two rectangles, whose areas add to $a^2 + 2ab + b^2$, and so the two quantities are equal. In the exercises, you will be asked to find a similar proof without words of the identity $(a + b)(c + d) = ac + ad + bc + bd$.

6.3   Criticism of the "without words" aspect

But let us not be dogmatic about the "without words" aspect of proofs without words. Proofs of any kind can be improved with insightful explanation, and there is little reason for a figure to stand entirely on its own. In fact, I know of numerous failed instances of proofs without words, where a clever diagram is too obscure by itself to constitute a mathematical argument. I regard these instances instead simply as poorly explained proofs. These clever diagrams could become part of a successful proof, if supplemented with proper explanation.

On my view, almost every proof without words is improved by a few well-chosen words, and for this reason I am critical of the *proof without words* label, when intended literally. Let us not make a false virtue of obscurity, taking the proof-without-words genre as a dogmatic, irritating insistence not to use words. Instead, let us understand the genre positively, as a celebration of the power of clever mathematical diagrams to convey mathematical ideas. Yes, let us give proofs that make use of insightful diagrams at their core, but let us also kindly explain them well, even if this means using a few words.

6.4 Triangular choices

Here is a particularly clever case. The triangular diagram is offered as a proof of the indicated combinatorial identity. The notation $\binom{n+1}{2}$ here denotes the number of ways to choose two objects from a set of $n + 1$ objects. Can you see how the picture proves the theorem?

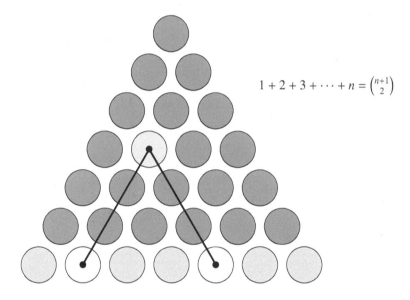

$$1 + 2 + 3 + \cdots + n = \binom{n+1}{2}$$

Aha! The idea is that there are precisely $1 + 2 + 3 + \cdots + n$ many blue circles on the first n rows of the figure, because the top row has 1 circle, the next row has 2, and so on, until row n. The key insight to make is that each blue circle is determined by specifying a pair of gold circles on the bottom row, as indicated by the connecting lines. Every pair of gold circles determines a different blue circle, and every blue circle is determined by such a pair of gold circles. Since this is a one-to-one correspondence, the number of blue circles is equal to the number of ways to choose two gold circles, and so $1 + 2 + \cdots + n = \binom{n+1}{2}$, as desired.

6.5   Further identities

Let us now consider several further questions concerning the identity of the previous problem.

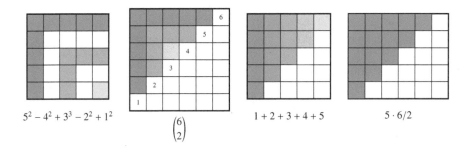

$5^2 - 4^2 + 3^3 - 2^2 + 1^2$ $\qquad \binom{6}{2} \qquad$ $1 + 2 + 3 + 4 + 5$ $\qquad$ $5 \cdot 6/2$

In the exercises, you will be asked to explain how these diagrams establish the identities:

$$\sum_{k=1}^{n}(-1)^{n-k}k^2 \;=\; \binom{n+1}{2} \;=\; \sum_{k=1}^{n}k \;=\; \frac{n(n+1)}{2}.$$

6.6   Sum of odd numbers

Next, consider this diagram as a proof that the sum of the first n odd numbers is n^2:

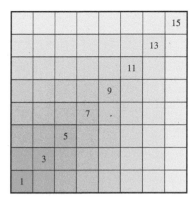

$$1 + 3 + 5 + \cdots + (2n - 1) = n^2$$

Can you see how the diagram proves the identity? The large square, with area n^2, is partitioned into bent sections of size 1, size 3, size 5, and so on, for the first n odd numbers. So the sum of those areas is equal to n^2, the total area. Do you prefer this proof to the induction argument we gave in theorem 23?

6.7 A Fibonacci identity

Consider next the traditional Fibonacci sequence, 0, 1, 1, 2, 3, 5, 8, 13, 21.... Each number is the sum of the previous two. If f_n denotes the nth term in this sequence, so that $f_0 = 0$, $f_1 = 1$, and $f_n + f_{n+1} = f_{n+2}$, then this diagram is offered as proof that $f_0^2 + f_1^2 + \cdots + f_n^2 = f_n f_{n+1}$:

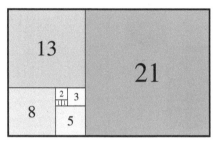

Do you see it? The squares arise from the Fibonacci numbers; their arrangement exactly ensures the Fibonacci recursion: $f_n + f_{n+1} = f_{n+2}$. Since the squares altogether at stage n correspond to a rectangle with sides f_n by f_{n+1}, we deduce the desired identity.

6.8 A sum of cubes

Consider next the following diagram, aiming to prove the cubic sum identity:

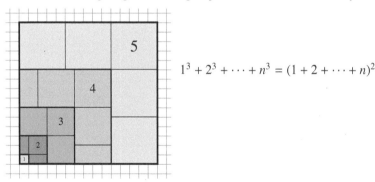

$$1^3 + 2^3 + \cdots + n^3 = (1 + 2 + \cdots + n)^2$$

Can you see how it works? Note that we have one tiny gold square of size 1. If we put the two red rectangles together, it makes in total two red squares of size 2. And similarly there are three orange squares of size 3, four squares of size 4, and five squares of size 5. So the sum of the areas of all the colored squares is the sum of cubes $1^3 + 2^3 + 3^3 + \cdots + n^3$ (why?). On the other hand, they are assembled into a large square, whose side length is $1 + 2 + 3 + \cdots + n$. So the two sides of the desired identity both represent the total area of the large square.

6.9   Another infinite series

This next diagram proves the indicated infinite sum. Can you see why?

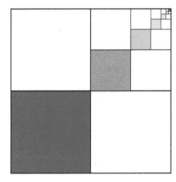

$$\sum_{n=1}^{\infty} \left(\frac{1}{2}\right)^{2n} = \frac{1}{3}$$

Aha! In the limit, exactly one-third of the square is shaded, since each shaded square is one of three squares of that size, and together they exhaust the unit square. The first square is $\frac{1}{2} \times \frac{1}{2}$, the next is $\frac{1}{4} \times \frac{1}{4}$, and so on. Since $(\frac{1}{2^n})^2 = (\frac{1}{2})^{2n}$, we get the sum of the shaded squares on the left-hand side, and the total area of $\frac{1}{3}$ on the right. So they are equal.

6.10   Area of a circle

Consider a disk of radius r, which has a circumference of $2\pi r$, by the definition of π. (I find it a deep theorem that all circles are similar, which is what causes π to be constant in this way.) Slice the disk like a pizza into a large number of very thin slices, and arrange them like this:

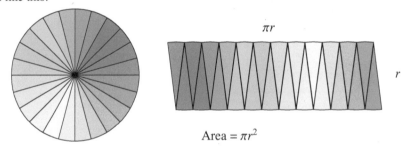

Area $= \pi r^2$

The rearranged pizza slices form an essentially rectangular shape of size πr by r, with half the circumference on top and half on bottom, and the thinner the slices we use, the more accurate these dimensions become, as accurate as we like. Thus, the area of a circle of radius r is πr^2.

6.11   Tiling with dominoes

Take a square grid with two opposite corners missing. Can you tile it with 2×1 dominoes? Place the dominoes, without overlapping, so as to completely cover the board.

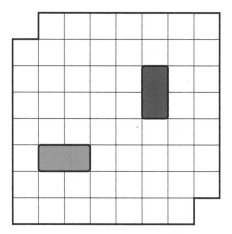

I have placed two red dominoes already, but please feel free to move them, if this would help you to find a way to tile the whole figure. Give it a try!

Interlude...

Did you find a tiling? Let me give you a hint. One way to argue begins by placing a chessboard pattern on the grid, as pictured here.

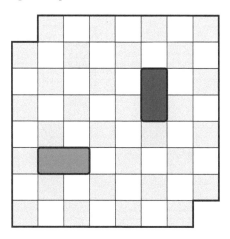

Can you see how to mount an argument with this chessboard pattern?

In fact, there can be no tiling of the omitted-corners board. The key observation to make with the chessboard pattern is that the omitted opposite corners would have had the same light color, and so there are more dark squares remaining than light squares. But since every domino covers adjacent squares, which have opposite colors, there can be no tiling.

The same argument shows more generally that we cannot omit squares of the same color and still expect to tile the resulting figure. But suppose we omit two squares of opposite color? Can we always find a tiling in this case? The answer is yes, and it does not matter which two squares we omit, as long as they have opposite chessboard colors. Here is a diagram containing the proof idea; you will explain it in the exercises.

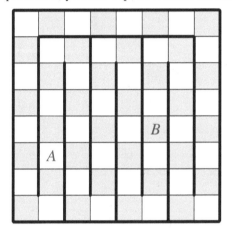

Here is another tiling problem. Remove a corner square from an ordinary chessboard; can you tile the remaining grid with long dominoes, that is, with dominoes of size 3×1?

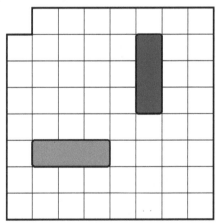

We need to cover 63 squares, a multiple of 3. But is there a tiling with 3×1 dominoes?

In fact, there is no tiling of this figure with long dominoes. To see this, however, the chessboard coloring of the squares is not so helpful. Instead, let us consider the 3-coloring of the squares pictured here:

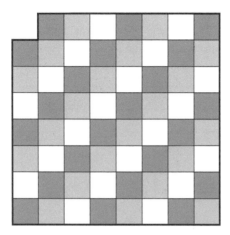

The thing to notice is that each 3×1 domino, no matter how it is placed on the grid, will cover exactly one square of each shade. But if you count, you will notice that there are only twenty white squares, while there are twenty-two dark blue squares. So there can be no tiling of the figure with long dominoes.

You might notice that the 3-coloring of the grid we just considered has the colorings on the next row shifted one space to the left. Suppose that we had instead made the coloring so as to shift one to the right, as shown here:

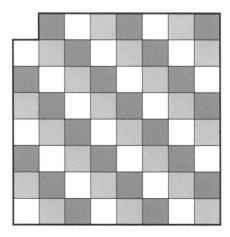

Would the argument still work? Can we reason just as before? You will be asked to answer in exercise 6.5.

6.12   How to lie with pictures

Let me conclude this chapter with a few examples showing how diagrams can sometimes mislead us in our reasoning. Consider the figure below. The colored tiles in the upper and lower figures are exactly the same shape and size. Yet, in the upper figure, they appear to fill out a triangle with base 13 and height 5, for a total area of 32.5, whereas in the lower figure, there is an extra white square, so the area of the colored tiles there would seem to be only 31.5.

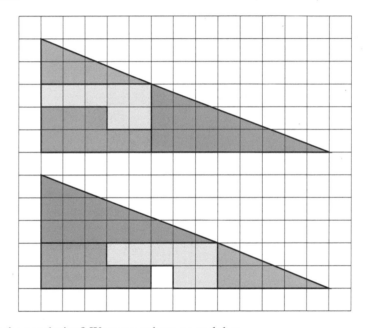

What is the conclusion? We seem to have proved that

$$31.5 \quad = \quad 32.5.$$

Amazing! What is wrong? Clearly, we should not trust this conclusion much. Please ponder this, until you understand what is going on. You will be asked to resolve the paradox in exercise 6.6.

Here is another mathematical gem, where a figure seems to lead us astray. Namely, I shall prove that every triangle is isosceles.

"Proposition" 47. *Every triangle is isosceles.*

Wait, um, . . . what? We know, of course, that this proposition is not true. We may easily construct many nonisosceles triangles. Nevertheless, the argument I shall give appears very solid, with every step seemingly correct. Can you spot any error?

"Proof" of proposition 47. Consider an arbitrary triangle △*ABC*.

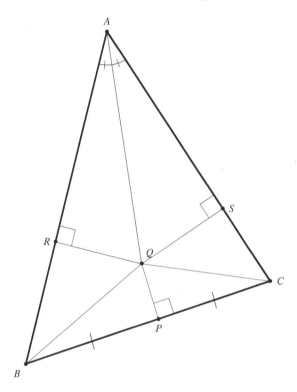

Let *Q* be the intersection of the angle bisector (blue) at ∠*A* and the perpendicular bisector (green) of *BC* at midpoint *P*. Drop perpendiculars from *Q* to *AB* at *R* and to *AC* at *S*. Because *P* is the midpoint of *BC* and *PQ* is perpendicular, we deduce that *BQ* ≅ *CQ* by the Pythagorean theorem. Since *AQ* is the angle bisector of ∠*A*, the right triangles *AQR* and *AQS* are similar, and since they share a hypotenuse, they are congruent. It follows that *AR* ≅ *AS* and also *QR* ≅ *QS*. Therefore, △*BQR* is congruent to △*CQS* by the hypotenuse-leg congruence theorem. So *RB* ≅ *SC*. And therefore,

$$AB \cong AR + RB \cong AS + SC \cong AC,$$

and so the triangle is isosceles, as desired. □

I once presented the proof argument above to an advanced mathematics research seminar at the University of California at Berkeley, as entertainment while we were waiting for the scheduled speaker. When the speaker arrived, however, a senior member of the audience insisted that we first get to the bottom of the puzzle, which had sent him into a fit of confusion. Are you confused? Why not?

Of course, we may also immediately deduce the following corollary.

"Corollary" 48. *Every triangle is equilateral.*

"Proof." The proof of proposition 47 proceeded from an arbitrary vertex *A* of triangle △*ABC*, and so the argument actually shows that each pair of adjacent sides is congruent. So it is equilateral.   □

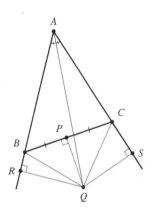

Perhaps someone might criticize the proof we gave for proposition 47 by saying that we do not necessarily know that the angle bisector at *A* intersects *BC* on that side of the midpoint *P*. Perhaps the intersection is on the other side, as in the diagram here. This would cause the point *Q* to be exterior to the triangle.

But the argument works just as easily for this case. Namely, we again let *Q* be the intersection of the angle bisector at ∠*A* with the perpendicular bisector of *BC* at midpoint *P*, and again drop the perpendiculars from *Q* to *R* and *S*. Again, we get *BQ* ≅ *CQ* by the Pythagorean theorem, using the green triangles. And again, we get △*ARQ* ≅ △*ASQ*
since these are similar triangles with the same hypotenuse. So again, we conclude that △*BQR* ≅ △*CQS* by the hypotenuse-leg congruence theorem. So we deduce that *AB* ≅ *AR* − *BR* ≅ *AS* − *CS* ≅ *AC*, and so the triangle is isosceles. □

Mathematical Habits

Use insightful diagrams and figures. Whenever possible, augment your mathematical arguments with graphical aids. Convey your mathematical ideas visually. Take the time to design informative pictures or diagrams. Use these graphical elements to illustrate and clarify your main argument, but not to replace it.

Recognize the limitations of figures and diagrams. When they do not or cannot represent the fully general case, figures and diagrams can sometimes lead to unwarranted conclusions, which hold for that particular case but not generally. So take care to consider whether your figures or diagrams might be suggesting particular rather than general features.

Exercises

6.1 Give an alternative proof by induction that $1 + 2 + \cdots + n = \binom{n+1}{2}$.

6.2 In the style of the binomial square figure, give a proof without words of the identity $(a+b)(c+d) = ac + ad + bc + bd$.

6.3 In the chessboard tiling problem, the grids that were pictured were 8×8, like a regular chessboard. But do the arguments depend on this? Does the result hold for any size square grid?

6.4 Using the figure appearing in the chapter, prove that if you delete two opposite-color squares from an ordinary 8×8 chessboard, you can tile it with 2×1 dominoes, no matter which two opposite-color squares are omitted. Does the argument generalize to any size square grid? What about rectangular grids?

6.5 Does the alternative 3-coloring of the chessboard, with one corner removed, still work with the argument that there is no tiling by 3×1 long dominoes?

6.6 Criticize the following "proof." Claim. $32.5 = 31.5$. Proof. Consider the diagram shown in the chapter. The upper triangle has base 13 and height 5, for an area of 32.5. The colored pieces can be rearranged as indicated to form the lower triangle, which has the same size, but it is missing exactly one square. So the area of the colored part is 31.5. Thus, $32.5 = 31.5$. $\square$

6.7 Explain how the following figure can be used to prove that $2\pi > 6$ and hence that $\pi > 3$.

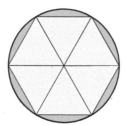

6.8 Explain how the figures appearing in the chapter establish the identities

$$\sum_{k=1}^{n}(-1)^{n-k}k^2 = \binom{n+1}{2} = \sum_{k=1}^{n} k = \frac{n(n+1)}{2}.$$

6.9 Find a proof without words in another collection that particularly inspires you, and present it to the class. Make sure to provide sufficient explanation that the rest of the class can understand and appreciate it; it is likely that a superior explanation will arise if you actually do use some words.

6.10 Criticize the proof of proposition 47. Is there any error in the reasoning?

Credits

Several of the examples in this chapter were adapted from answers to the MathOverflow question posed by Mariano Suárez-Álvarez (2009), and many also are part of mathematical folklore. The triangular choices example is due to Larson (1985); see also Suárez-Álvarez (2009). The proof that dominoes can tile the chessboard with two opposite-color squares omitted is due to Ralph E. Gomory; see Shreevatsa (2010). The series $\sum_n (\frac{1}{2})^{2n} = \frac{1}{3}$ was reportedly known to Archimedes and used in his *Quadrature of the Parabola*; see Silagadze (2014). The sum-of-cubes example is due to Brian Sears; see Pritchard (2010). I heard the 3×1 long domino tiling example from Kevin O'Bryant, who reportedly learned it at math events in Hungary. The proof that $31.5 = 32.5$ is reportedly due to New York City amateur magician Paul Curry in 1953; see O'Connor (2010). The proof that all triangles are isosceles is reportedly due originally to W. W. Rouse Ball, *Mathematical Recreations and Essays* (1892). It was Robert Solovay, at the research seminar, who insisted on getting to the bottom of it—and he did indeed.

7 Theory of Games

Nearly every mathematician likes a good game. Let's play! There is plenty of mathematical analysis to be undertaken. We shall start with a few fun games and their winning strategies before moving on to develop some of the general theory in the logic of games, ultimately proving the fundamental theorem of finite games, due to Ernst Zermelo, which shows that every finite game has a winning strategy for one of the players or a drawing strategy for both players.

7.1 Twenty-One

Consider first the game of Twenty-One, in which two players cooperate to count to twenty-one, with each player saying either the next one, two, or three numbers, starting at one. Whoever says "twenty-one" is the winner. Perhaps our game proceeds like this.

You One, two.

Me Three, four, five.

You Six, seven, eight.

Me Nine.

You Ten, eleven.

Me Twelve, thirteen.

You Fourteen.

And so on. How shall I reply next? Who will be able to win by saying "twenty-one"? Kindly find a partner and play the game a few times. Get a good feel for the game. Can you find a winning strategy? This is a game that you can likely figure out.

Interlude...

Did you find a winning strategy? Perhaps one might hope to discover how to win by working backward from the winning condition. Imagine what number you must say on your penultimate turn, the number from which your opponent cannot win, but you will be able to win right after that. This would be the number 17, of course, since if you say 17, then your opponent cannot quite reach 21, but whether your opponent ends on 18, 19, or 20, you will then be able to reach 21 and thereby win. By working further back from this, we identify the winning strategy.

Theorem 49. *The first player has a winning strategy in the game of Twenty-One. The winning strategy is to end each turn with one of the numbers on this list:*

$$1, \ 5, \ 9, \ 13, \ 17, \ 21.$$

Proof. That is, the first player should always stop counting on each turn exactly at one of the numbers on the list. Notice that the first player can certainly start with a number on the list, simply by saying the number one. And since the numbers on the list are spaced four apart, whenever the first player has stopped at a number on the list, the opponent cannot get up to the next number on the list, since the opponent can add only at most three numbers. The key observation to make is that, because the opponent must add at least one on each turn, it follows that the first player will thereby come within striking range of the next number on the list, enabled to climb to it on the next turn. So the first player will ultimately be able to say "twenty-one" and therefore win.   □

I would like to emphasize the uniqueness aspect of the winning strategy: we had referred to *the* winning strategy, rather than *a* winning strategy. What I claim is that the strategy we identified is in fact the only winning strategy, for if a player should ever fail to follow the strategy, by saying a number not on the list, then the argument of the proof shows that the opponent can immediately seize control of the game and take up the strategy simply by climbing to the next number on the list.

The game of Twenty-One generalizes naturally to many other versions, where we adopt a different target number or allow a different step size at each turn. For example, perhaps we aim to count up to thirty-five or some other number; or perhaps we allow each player to announce up to five numbers on each turn, or perhaps only two. So we really have an infinite parameterized scheme of games here to analyze. Let $G_{n,s}$ denote the version of the game with target n and allowed step size s, meaning that the two players cooperate to count to n, starting with one and each adding up to the next s numbers on each turn; whoever gets to n wins. In this notation, the original game of Twenty-One is $G_{21,3}$, with a target of 21 and step size 3. Kindly play a few rounds of the game $G_{31,5}$ with your partner. Can you find a winning strategy? Can you generalize the idea behind the proof of theorem 49?

Interlude...

Indeed, there is a general strategy for all these games.

Theorem 50. *In the game $G_{n,s}$ for any positive integers n, s, the first player has a winning strategy if and only if n is not a multiple of s + 1, and otherwise the second player has a winning strategy. In each case, the winning strategy is to end your turn with exactly the numbers with the same remainder as n modulo s + 1.*

In the game $G_{31,5}$, for example, the theorem says that the first player will win by ending each announcement on one of these numbers, the numbers congruent to 31 modulo 6:

$$1, \quad 7, \quad 13, \quad 19, \quad 25, \quad 31.$$

Proof of theorem 50. Let r be the remainder of n modulo $s + 1$. Since the numbers with remainder r modulo $s + 1$ are spaced $s + 1$ apart, when a player says such a number, the opponent cannot get to the next such number with remainder r. But after any play by the opponent, which must add at least one, then the original player will be able to get to the next such number.

So whichever player can say a number with remainder r modulo $s + 1$ will be able to continue to say such numbers all the way up to n. Since $r \le s$, the first player will be able to stop at r, unless $r = 0$. So the first player will have a winning strategy, unless n is a multiple of $s + 1$. If n is a multiple of $s + 1$, then the first player will say some number at most s on their first turn, and the second player will be able to reach $s + 1$, thereby getting on the winning list. So in this case, the second player has a winning strategy. $\square$

7.2 Buckets of Fish

In chapter 4, we introduced the game Buckets of Fish, which is played with finitely many buckets in a row on the beach, each containing some finite number of fish, and a large supply of additional fish available nearby, fresh off the boats. On your turn, you may take one fish from one of the buckets and then add as many fish as you want to each of the buckets to the left. Whoever takes the very last fish from the buckets wins. Since potentially a large number of fish could be added to the buckets, which of course extends the length of play each time, it might seem that the players could conspire to make an infinitely long play, which never ends and for which there is no winner. Nevertheless, in theorem 32, we proved that this is impossible. Every play of the game ends in finitely many moves with a winner. But what is the strategy? Say you are faced with buckets having fish in the following amounts:

$$4 \quad 5 \quad 2 \quad 0 \quad 7 \quad 4$$

What is your winning move? Please give it some thought before reading further!

Interlude...

Welcome back. Did you find a winning strategy? I was surprised to find the following winning strategy, which was much simpler than I had expected at first would be possible.

Theorem 51. *The winning strategy in the game Buckets of Fish is to play so as to ensure that every bucket has an even number of fish.*

Proof. Notice first that in the case that there is only one bucket, then if it contains an even number of fish, the second player can win, since the first player will necessarily make it odd, and then the second player will make it even again. Thus, it will be the second player who will make it zero, winning the game. So in the trivial instance of the game with only one bucket, the player who can make the bucket even will be the winner.

Next, notice that if you play so as to give your opponent an even number of fish in every bucket, then whatever move your opponent makes will result in an odd number of fish in the bucket from which he or she takes a fish (and possibly also an odd number of fish in some of the earlier buckets as well, if your opponent happens to add an odd number of fish to some of them). So if you give your opponent an all-even position, then your opponent cannot give you back an all-even position.

Finally, notice that if you are faced with a position that is not all even, then you can simply take a fish from the rightmost odd bucket, thereby making it even, and add fish if necessary to the earlier buckets so as to make them all even. In this way, you can turn any position that is not all even into an all-even position in one move. By following this strategy, a player will ensure that he or she will take the last fish, since the winning move is to make the all-zero position, which is an all-even position, and we have already observed that the opponent cannot produce an all-even position.   □

In the particular position of the game mentioned before the theorem, therefore, the winning move is to take a fish from the bucket with seven fish and add an odd number of fish to the bucket with five fish, thereby producing an all-even position.

7.3   The game of Nim

The game of Nim is a mathematician's delight. The winning strategy is fundamentally mathematical, with just the right level of complexity so that a person can enjoyably implement it in actual games, but difficult enough so that an opponent who does not know the strategy is unlikely to play reliably in accordance with it. Those in the know can therefore usually expect to win—nearly every time—against those who do not know the strategy, even when starting from a random or losing position. I have taught young children the strategy, who then go on to defeat adults systematically. What fun!

Nim is a two-player game. To play, set up finitely many piles of coins in front of you, and decide who will go first. Any starting pattern is fine ($1, 3, 5, 7$ is common). On each turn, a player selects a pile and removes one or more coins from that pile—taking the whole pile is fine. The player who takes the very last coin wins.

In the instance below, we started with piles of height $1, 3, 4, 5, 7, 4, 2, 1$, and after the first player removed three coins from the third pile, the second player removed the entire sixth pile.

Actually, we shall see by the end of the chapter that both of these moves were terrible! How should they have played instead? (Come back and answer by the end!)

Mathematicians have the habit, when confronted with a difficult problem, to consider in detail a very simple case or boundary instance of the problem. Because of their simplicity, such instances sometimes make plain aspects of a problem that will help much more generally. For example, would it be fruitful to consider a Nim position with only one pile? Well, it is clear that the first player can win: take the entire pile, so perhaps that was not so fruitful. How about a game of Nim with two piles? This simple case will turn out to be very fruitful indeed. Consider a Nim position with two piles.

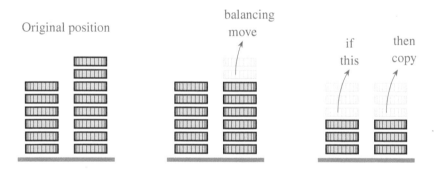

If they have different heights, then the first player can play so as to balance them, and the key observation is that every subsequent move will admit a copying reply to rebalance. In this way, the always-balance strategy becomes a winning strategy, because if one can always copy, then in particular one will have a move whenever the opponent did, and so the opponent will never take the last coin. This observation provides a complete analysis for the case of Nim with two piles: if they are unbalanced, then the first player has a winning strategy to always (re)balance them; if they are already balanced, then the second player can implement the balancing strategy.

Theorem 52. *In any Nim position consisting of pairs of same-height piles, the second player has a winning strategy.*

Proof. If the Nim position consists of pairs of same-height piles, then the winning strategy for the second player is simply to copy the opponent's move on the counterpart pile pair.

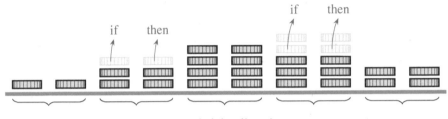

same-height pile pairs

Since any move on a pair-balanced position will admit such a rebalancing move, the second player will be the one to make the last move and thereby win.   □

The deeper game-theoretic observation to make here is that, when analyzing a given Nim position, we can ignore pairs of same-height piles—in effect, they cancel each other out—since any move on them can be copied, and those moves do not affect the rest of the position. The balanced-piles idea leads immediately to a winning strategy in the case that all piles have height 1 or 2 at most. Let us say that such a position is *balanced* if the number of piles of height 1 and of height 2 are each even numbers.

Theorem 53. *In any Nim position with piles all of height at most 2, one of the players has a winning strategy.*

Proof. If the position is already balanced, then our previous observation implies that the second player has a winning strategy. For the remaining case, suppose the position is not balanced but that all piles have height at most 2. I claim that there is a balancing move. Since the position is not balanced, either there is an odd number of 1-piles, an odd number of 2-piles, or both. If only one type of pile has an odd number of instances, then the first player can balance the position simply by removing an entire pile. If both pile types have an odd number of instances, then the first player can balance the position by removing one coin from a 2-pile, thereby in effect adding a 1-pile while destroying a 2-pile.   □

To summarize, we have shown that if a position is balanced with all piles of height at most 2, then the second player has a winning strategy, the balancing strategy; and if the original position is unbalanced, then the first player can balance it and thereby adopt the balancing strategy. A clever generalization of the concept of "balanced," it turns out, will enable us to handle the general case and thereby find a complete winning strategy for Nim. However, we shall *not* define that a position is balanced if and only if it can be partitioned into pairs of equal-height piles, which might seem initially to be a sound generalization of the "balanced" idea. Rather, the more useful generalization is subtler.

Definition 54. A Nim position is said to be *balanced* if, when one represents each pile height as a sum of distinct powers of 2, as in theorem 30 (see chapter 4), then for each individual power of 2 that arises, the total number of occurrences of it as a summand in this way is even.

Equivalently, as the reader will verify in exercise 7.4, a Nim position is balanced if and only if there are overall even numbers of 1s in each digit place, when we represent the pile heights in binary notation. The traditional Nim starting position, played at bars, has piles of heights $1, 3, 5, 7$, and this is balanced, because in the powers-of-2 representations, as indicated, we have two 4s, two 2s, and four 1s, an even number in each case.

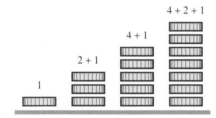

The winning nature of the Nim balancing strategy ultimately amounts to the following:

Lemma 55.

1. *If a Nim position is unbalanced, then there is a balancing move.*
2. *Any move on a balanced Nim position will unbalance it.*

Proof. For statement (1), suppose that we have an unbalanced Nim position. Represent the pile heights as sums of distinct powers of 2, and let 2^k be the largest unbalanced power of 2 that occurs. Consider any pile in which this largest power, 2^k, occurs as a summand. We will move on this pile so as to make the position balanced. We shall concern ourselves only with the 2^k and smaller summands of this pile. Since $1 + 2 + \cdots + 2^{k-1} = 2^k - 1$ by theorem 24 (see chapter 4), we can in principle remove the 2^k and leave any desired collection of smaller powers of 2 for this pile, since even leaving all the smaller powers of 2 only adds up to $2^k - 1$. The balancing move, therefore, will be to reduce the pile to a height that preserves all the previous larger powers of 2 and that contains exactly the smaller power-of-2 summands 2^r for $r < k$ needed in order to balance them. That is, if 2^r occurs at other piles an odd number of times, then we include a summand of 2^r in this pile to make it even overall. In this way, all the powers of 2 are balanced in the new position.

For statement (2), suppose now that we have a balanced position, and we move on it. This means that we reduce one of the piles from having height n to having some height m less than n. Since $m < n$, these two numbers do not have exactly the same powers of 2 in their representations as sums of distinct powers of 2, and so there must be some power of 2—let us denote it 2^k—that occurs in n or m but not both. So the move from n to m must change the even-odd parity of the number of occurrences of 2^k as a summand in the position overall. Since it was originally balanced, therefore, it must now be unbalanced.   □

Let us refer to the *balancing* strategy, which is the strategy that always plays a balancing move on any unbalanced position.

Theorem 56. *The balancing strategy is a winning strategy in the game of Nim. If the initial position is unbalanced, then the first player can win by always playing balancing moves. If the initial position is already balanced, then the second player can win by always rebalancing.*

Proof. By lemma 55, any player faced with an unbalanced position has a balancing move, and any subsequent move by the opponent will unbalance it again. So by induction, a player faced with an unbalanced position can continue to always give balanced positions to the opponent, while in this case the opponent will always give unbalanced positions back. Since the number of coins overall is steadily decreasing, the game will eventually end, and the winning move is a move to the all-zero position, which is balanced. So the player who is always giving balancing positions will thereby ensure that she is the winner.   □

The reader is invited to return to the initial Nim example earlier in this chapter in order to explain how the first player should have played.

Misère Nim.

In what is called the misère form of Nim, one wants *not* to take the last coin. The reader will discover and verify the winning strategy for misère Nim in the exercises.

Transfinite Nim.

Those readers who are familiar with the transfinite ordinals will enjoy considering transfinite Nim, where the piles can have transfinite ordinal height. So we have finitely many piles of ordinal height, perhaps infinite, and a move consists of making any one pile strictly shorter. Since there are no infinite descending sequence of ordinals, the game will terminate in finitely many moves, and the winner is whoever removes the last coin. Because ordinals have unique representations as sums of (decreasing) distinct ordinal powers of 2, the analogue of theorem 30 works for ordinals, and this is sufficient to prove that the balancing strategy still works in transfinite Nim.

7.4   The Gold Coin game

Consider next the Gold Coin game, a two-player moving-coin game played with a gold coin and some pennies. Each player strives to get the gold coin.

The game begins with the coins on a track, and the gold coin is farthest to the right. On each turn, a player may either (1) move any coin to the left by one or more spaces but without jumping over another coin, or (2) take the leftmost coin into his or her possession. Whoever takes the gold coin is the winner. In the diagram above, one of the players has just moved the gold coin three spaces, which we will see is a winning move. What is the winning strategy? I would recommend playing the game with a partner several times in order to gain familiarity with it.

It turns out that the Gold Coin game can be successfully analyzed by means of a reduction to the game of Nim, which will lead to a winning strategy in the Gold Coin game. In order to carry out this reduction, we associate with each position in the Gold Coin game a certain Nim position as follows.

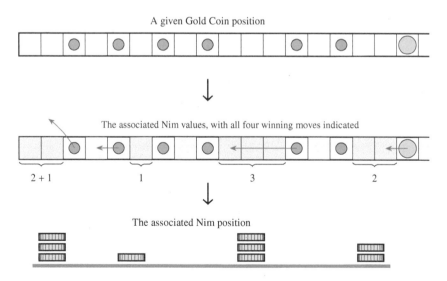

Beginning at the right with the gold coin, we take the coins in successive pairs, and for each such pair of coins, we think of the interval between them as a Nim pile (orange); we ignore the intervals between these pairs (blue); and if the leftmost coin has no partner, then the corresponding Nim pile is the number of open squares to the left of it, plus one. In this way, every position in the Gold Coin game is associated with a Nim position.

Let us define that a Gold Coin game position is *balanced* if the corresponding Nim position is balanced. The Gold Coin position indicated in the diagram above, for example, is associated with the Nim position 3, 1, 3, 2, which is not balanced. But there are a variety of balancing moves, and the Gold Coin moves that lead to balanced Nim positions are indicated with the red arrows in the figure. One subtle point to notice is that moving via the second arrow leads to a Gold Coin position associated with the Nim position 3, 2, 3, 2, which is not reachable from the Nim position 3, 1, 3, 2 associated with the original Gold Coin position, since one of the piles became higher. So the legal Gold Coin moves do not always correspond to legal Nim moves in the associated position. Nevertheless, the association is fruitful, because it respects the balanced-position concept.

Lemma 57.

1. *If a Gold Coin game position is not balanced, then there is a balancing move.*
2. *If a Gold Coin position is balanced, then every move will unbalance it.*

Proof. If a Gold Coin game position is not balanced, then the associated Nim position is not balanced. By lemma 55, there is a balancing Nim move on that position. Every valid Nim move on the associated Nim position can be realized by means of a valid Gold Coin move, by moving the corresponding coin for that pile. So there is a balancing move in the Gold Coin game. If alternatively the Gold Coin game position is balanced, then the key observation is that every move on a Gold Coin position affects at most one of the associated Nim piles, since every coin is part of at most one such pile. It follows that any move in the Gold Coin game position will affect only one pile in the associated Nim position and must therefore result in an unbalanced Nim position.   □

While the balancing player can always play Gold Coin moves that correspond directly to moves in the associated Nim game, nevertheless the opponent might play in a way that does not correspond to any legal Nim move, by moving or taking the left coin in a coin pair, for example, which would correspond to making one of the Nim piles larger. This is fine for the balancing player, however, since that move affects only one pile and therefore leads to an unbalanced associated Nim position. So the balancing player can simply continue to play balancing moves. And although the associated Nim piles can sometimes grow, it is clear that the Gold Coin game will terminate in finitely many moves, since the total distance of the coins to the left edge is finite and strictly decreasing with each move.

Theorem 58. *The balancing strategy is a winning strategy in the Gold Coin game.*

Proof. If one player gives a balanced position to the other in the Gold Coin game, then by lemma 57, the opponent must unbalance it, and so the original player can balance it again. By induction, therefore, the original player can continue always to give balanced positions. This is a winning strategy, since the winning move is a balancing move. If the original position is unbalanced, then the first player can adopt this strategy by playing a balancing move. If the original position is balanced already, then the second player can adopt the balancing strategy since the first player will unbalance it. □

Those readers who go on to study the theory of games more intensely will eventually find the Sprague-Grundy theorem, which asserts that every impartial game can be translated to Nim in the way we have done for the Gold Coin game. So this method, it turns out, is extremely general.

7.5 Chomp

Let us now consider another fun game, called Chomp, which I shall use to illustrate an interesting point on the difference between proving that something exists and being able to exhibit it. Consider an $n \times m$ rectangular chocolate bar, consisting of squares.

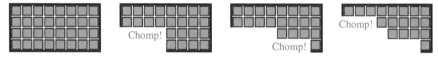

Players "chomp" the bar from the lower left, by selecting a particular square in the bar and removing it along with the rest of the bar to the southwest of it. The player who chomps the last square at the upper right loses.

Let us make some easy observations. First, it is clearly losing to leave a single row or column of chocolate (unless it is only the last square), since the opponent could chomp all but the last square like this:

a winning chomp

Second, it is winning to leave balanced "wings," as pictured here.

This is like a Nim position with two balanced piles—any move on one side can be matched with a move on the other side, thereby forcing your opponent to take the final square. Using this fact, the reader will prove in exercise 7.12 that there is an explicit winning strategy in Chomp on any square.

The reader may be surprised to learn that the game of Chomp, although very simple, is in a sense an open mathematical question—no one has yet described a feasible winning strategy for the general case! We can, of course, describe strategies for various special instances of the game; what we lack is a feasible general solution that works on all size chocolate bars. It will follow from the proof of theorem 60 that there is an explicit winning strategy, obtained as in that proof from the game tree and the back-propagation method, but in practice, this method is not feasible, since searching the game tree takes a long time. Meanwhile, I find it fascinating that, although we do not know the general winning strategy for Chomp, nevertheless we do know that it must be the first player who has a winning strategy in any nontrivial instance of the game. We can give a pure-existence proof that there is winning strategy for the first player, even though we can prove essentially nothing about this winning strategy or what moves it might entail.

Theorem 59. *In any nontrivial instance of the game of Chomp, the first player has a winning strategy.*

Proof. We shall prove this theorem by means of a *strategy-stealing* argument, relying on the fundamental theorem of finite games, which says that in any finite game, one of the players has a winning strategy. That theorem will be proved later in this chapter (theorem 60).

Consider a nontrivial instance of the game of Chomp, nontrivial in that it is not the 1×1 instance of the game. In order to show that the first player has a winning strategy, it suffices by the fundamental theorem of finite games to show that the second player can have no winning strategy. So, suppose toward contradiction that the second player has a winning strategy σ, which tells the second player how to chomp, given the sequence of play so far. Using σ, I shall describe a strategy for the first player. What the first player will do is pretend to be the second player, by inventing an imaginary first move for his opponent and then using the strategy σ from that point on. Specifically, the actual first player will pretend that his opponent has gone first by chomping the 1×1 square, the smallest-possible chomp. The strategy σ will give a move in reply, and the important thing to notice is that however σ replies, it will be with a chomp that contains the 1×1 chomp. That is, any move made after the 1×1 chomp results in a position that could have been reached directly on the first move. Therefore, however σ replies to the 1×1 chomp will be a move that can legally be played as a first move. Let us direct the actual first player to make that move as his first move, and henceforth he pretends to be playing the game in which he is the second player and his opponent had begun with the 1×1 move. Since σ was a winning strategy for the second player, this means that the actual first player will win the imaginary game, which has all the same moves as the actual game, except for the imaginary first move. In particular, the first player will win, because he was the second player in the imaginary game.   □

7.6 Games of perfect information

Let us now consider games somewhat more abstractly in order to develop a little of the theory of the logic of games. What is a game, really? In the kind of game known as a two-player game of perfect information, there are two players, who take turns making moves, and a play of the game consists of a sequence of such moves, played until the outcome of this particular game play is known. The phrase *perfect information* is meant to suggest the idea that both players play from a position of complete knowledge about the state of the game; neither holds any hidden knowledge or secret powers concerning how the play could proceed. A *position* in the game is simply a legal sequence of moves; we imagine how play might proceed from that position, after those moves had been played. We shall include in this the empty sequence, which corresponds to the starting position of the game, as well as the positions arising right when a winner has become known. The collection of positions for a given game has the structure of a tree, known as the *game tree*.

A fragment of the game tree for the familiar game of tic-tac-toe is shown in figure 7.1. The game begins with an empty 3×3 grid, and player X may place her mark on any of the nine squares, followed by player O and then X again in turn. Each player is trying to make tic-tac-toe, or three in a row, and the first player to do so wins. The initially empty board is shown, along with the nine possible first moves for player X. In the full game tree, each of these nine boards leads in turn to eight possible replies by O, for 72 possible boards after two moves, each of which admits seven possible follow-up moves by X, for 504 possible boards after three moves, and so on. The entire game tree is quite large. (One can cut down the game tree a little by observing that there are essentially only three first moves, not nine, namely, the corner move, side move and middle move; by symmetry, it does not matter which particular corner or side is chosen—and there are also a few symmetries still after two moves and sometimes more, but the tree is still very large.) Here, we have pictured just part of the game tree, showing the choices that arise during the course of a particular play of the game, one in which X ultimately wins. The game of tic-tac-toe, like chess, is a bit different from other games that we shall consider in that some plays of tic-tac-toe and chess lead to a *draw* situation, where the game is over but neither player has won. In other common kinds of games, every outcome is a win for exactly one of the players.

Any two-player game of perfect information has an associated game tree, which contains in a sense all the relevant information about the game, delimiting the space of legal moves and indicating who wins any particular play (or whether it is a draw). Indeed, every finite tree can be viewed as a game tree, if one should merely specify of the terminal nodes the winner of that particular play. The game begins at the top node of the tree, and each player in turn selects a child node of the current node, until a terminal node is reached and the outcome becomes known. We therefore define that a *finite game* consists of a finite tree, whose terminal nodes are labeled as a win for either the first or the second player, or as a draw.

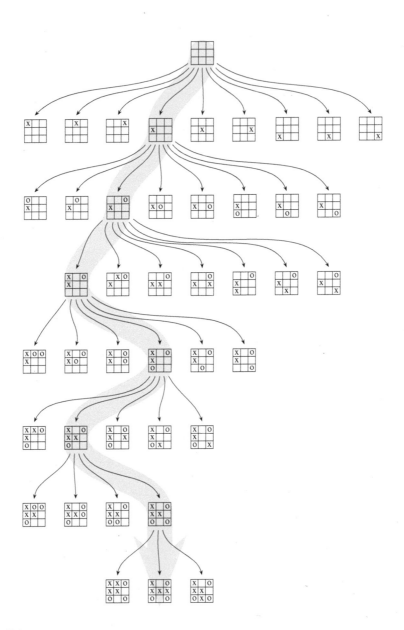

Figure 7.1
A tiny fragment of the tic-tac-toe game tree, illustrating a particular winning play for *X*

Here is the game tree of a game to be played by Alice and Bob. Play starts at the top; Alice makes the first choice, then Bob, and then Alice again. The winner is indicated in the resulting final node. Who can force a win?

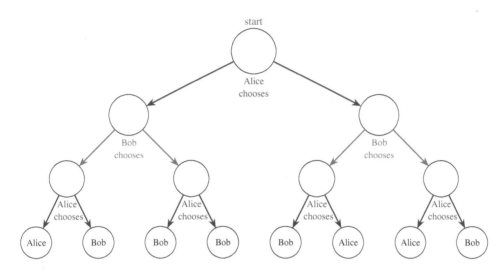

A *strategy* in such a game is a function that tells a particular player, from any node, which child node to select from that position. A play of the game accords with a strategy for a particular player if at every node when it was that player's turn, he or she did indeed select the child node that the strategy specified. A strategy is a *winning* strategy for a player if every play that accords with that strategy results in a win for that player. A strategy is a *drawing* strategy for a player if every play that accords with that strategy results in either a win for that player or a draw.

7.7   The fundamental theorem of finite games

We are now ready to prove the fundamental theorem in the theory of finite games, due to Zermelo in 1913. I shall give three independent proofs for this central game-theoretic result. First, consider the case of games without draws.

Theorem 60 (Fundamental theorem of finite games). *In any finite two-player game of perfect information, without draws, one of the players has a winning strategy.*

First proof. This proof uses the *back-propagation* method, a form of backward induction. Consider the underlying game tree T. We intend to label the nodes of the game tree with the player who can win if play should start from that position. The terminal nodes are already labeled. Working backward, consider a node u all of whose children are already labeled. If it is Alice's turn to play and there is a child node labeled Alice, then Alice could play from u to that node, and so we also label u for Alice. If all the children of u are labeled Bob, then the situation is hopeless for Alice from position u, and we label u for Bob. Similarly, if it is Bob's turn and he may move from u to a node labeled Bob, then we label u for Bob, and otherwise for Alice. By this procedure, every node in the tree gets labeled either Alice or Bob. Each player wants to be on nodes with their label.

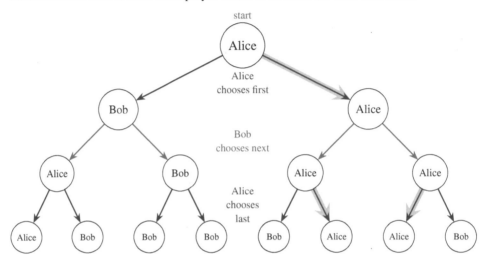

Our earlier game tree will become labeled as above. These labels, I claim, provide a winning strategy: just stay on the nodes with your label (winning moves indicated in purple for Alice in the figure). If the game play is currently on a node with your label, then if it is your turn, you can preserve that, by the way we defined the labels; and if it is your opponent's turn, all of their moves will still have your label. So whichever player gets his or her label on the initial position will have a winning strategy in the game.   □

Second proof. Let W be the set of nodes in the game tree from which player 1 has a winning strategy. If this includes the initial game position (the top node), then we are done. Otherwise, I claim that player 2 has a winning strategy, which is to avoid the nodes of W. If play is currently at a node u not in W, and it is player 1's turn, then no child can be in W, or else player 1 would have had a winning strategy from u. And if it is player 2's turn from u, then there must be a move to a node not in W, or else player 1 would have had a winning strategy from u, by amalgamating the strategies from the children of u. So player 2 can avoid the nodes of W and therefore avoid the terminal nodes where player 1 has won. So in this case, player 2 has a winning strategy.   □

The third proof will explain the sense in which the fundamental theorem of finite games amounts to the De Morgan law of quantifier logic, which asserts that $\neg \forall x \, \varphi(x)$ is logically equivalent to $\exists x \, \neg \varphi(x)$. For example, not every page in this book is scribbled upon just in case there is some page that is not scribbled upon. The quantifier symbol $\forall$ here is read as "for all," and the symbol $\exists$ as "there exists."

Third proof. Consider now any finite game, and for simplicity, let us assume that all plays of the game have the same number of moves n for each player (we could invent an equivalent game with imaginary moves so as to achieve this assumption without loss of generality). The assertion "player 2 has a winning strategy" can be expressed in logical symbols as the assertion

$$\forall x_1 \exists y_1 \forall x_2 \exists y_2 \cdots \forall x_n \exists y_n \text{ play } \vec{x}\vec{y} \text{ is a win for player 2,}$$

which simply asserts that for every move for player 1, there is a reply for player 2 and so on iteratively in such a way that the game is a win for player 2. The assertion that player 2 does *not* have a winning strategy is the negation of this statement. By the De Morgan law for negation of quantifiers, we know that $\neg\forall x$ is the same as $\exists x\neg$, and so we may push the negation all the way through the quantifiers, switching each of them to the dual quantifier. So the assertion "player 2 does not have a winning strategy" is equivalent to

$$\neg\forall x_1 \exists y_1 \forall x_2 \exists y_2 \cdots \forall x_n \exists y_n \; \vec{x}\vec{y} \text{ is a win for player 2,}$$

which is equivalent to

$$\exists x_1 \forall y_1 \exists x_2 \forall y_2 \cdots \exists x_n \forall y_n \; \vec{x}\vec{y} \text{ is a win for player 1.}$$

And this is precisely what it means for player 1 to have a winning strategy. So we have shown that if player 2 does not have a winning strategy, then player 1 does, and so one of them must have a winning strategy. □

Corollary 61 (Fundamental theorem of finite games with draws). *In any finite two-player game of perfect information, possibly with draws, one of the players has a winning strategy or both players have drawing strategies.*

Proof. Consider any such game G, and let us refer to the players as black and white, as in chess. Consider the related game G_W, where everything is played as in G, but now we count a drawn outcome as a win for white. By the fundamental theorem, one of the players has a winning strategy in G_W. If black has a winning strategy in G_W, then this strategy is also a winning strategy in G. Otherwise, white has a winning strategy in G_W, which is a drawing strategy in G. Similarly, if we consider the game G_B, where now a draw counts as a win for black, then by the fundamental theorem, one of the players has a winning strategy. If white has a winning strategy in G_B, then this is a winning strategy in G. Otherwise, black has a winning strategy in G_B, which amounts to a drawing strategy in G. So either one of the players has a winning strategy in G, or else both players have drawing strategies. □

Corollary 62. *In the game of chess, one of the players has a winning strategy or both players have drawing strategies.*

Proof. According to the standard tournament rules, chess is a finite two-player game of perfect information with the possibility of draws. So the claim made in this corollary is an instance of corollary 61.   □

Consider next a fun conundrum, the Hypergame paradox. Let us say that a game is *finite* if every play of the game ends in finitely many moves (this is more general than our earlier definition, which required the entire game tree to be finite, because there are some games that have an infinite game tree even though every play of the game is finite). Let me describe a new game, called Hypergame, in which the first player begins by selecting a particular finite game, and then play proceeds as in that game. Perhaps the first player selects the game arising from a particular Nim position, or a particular size of Chomp or whatever, and then play proceeds as in that game. Because the first player is required to select a finite game and then play proceeds as in that game, it seems clear that every play of Hypergame will end in finitely many moves, right? In other words, it seems that Hypergame is a finite game. In this case, it would seem to be allowed for player 1 to select Hypergame itself for the first move as the finite game to be played. And now play proceeds as in Hypergame, and so player 2 can make the first move in that game. But let us imagine that player 2 does so by also selecting Hypergame, which we have said is a finite game. After this, player 1 must move in that game and can again play Hypergame as the finite game to be played, followed by Hypergame by player 2 again, and so on, repeating forever. So we have found a play of the game that does not end in finitely many moves, contradicting our earlier observation that Hypergame was a finite game. Can you explain the paradox?

Let me conclude the chapter with a remark on taxonomy. There are three overlapping, interrelated but distinct research areas concerned with games. *Game theory*, closely related to decision theory, is concerned with game-theoretic equilibria, such as the Nash equilibria, and questions in economics, political science, social science, auction theory, and voting theory. The *theory of games*, in contrast, which is also known as *combinatorial game theory*, is the study of actual games, such as chess, go, Nim, and so on, and gave rise to the mathematical analysis of these games, such as we have given in this chapter, as well as to the Conway game numbers, which are closely connected with the surreal numbers. Finally, the *logic of games* is concerned with the underlying logic of game processes and analysis. This chapter was mainly about the theory of games and the logic of games, rather than game theory.

Mathematical Habits

Express key ideas several times in different ways. Say the same thing twice but differently. Explain your idea plainly at first, and then explain it in a more sophisticated way, with helpful details, if these give a fuller account, and summarize what you have done.

Take trivial and boundary cases seriously. Consider the trivial and boundary-case instances of your claim, and take them seriously. Check your claim in especially simple instances or instances that are trivial in part. Such cases can often reveal aspects of your concept that apply much more generally but that might be hidden in more complex situations. This advice is part of using words precisely and accurately, since an extreme or trivial case of a concept is still an instance of it. One can often refute an overly general test conjecture simply by observing that it fails in an extreme or trivial case. Mathematicians often pass by habit immediately to the extreme cases, because of the insights to which they can lead.

Exercises

7.1 True or false: in the game Twenty-One, making a mistake early on does not matter so much, since you can recover by the end by adopting the winning strategy. [Make a precise mathematical claim about this and prove it.]

7.2 Prove that the winning strategy we identified in theorem 50 for $G_{n,s}$ is unique; deviating from it is a losing move.

7.3 Consider the game Fifteen, in which two players take turns selecting numbers from 1 to 9, each selected at most once; whoever first has three numbers adding to fifteen wins (it must be three numbers adding to fifteen, not two). Using the following matrix, prove that this game is isomorphic to tic-tac-toe. Conclude that there is no winning strategy.

$$\begin{pmatrix} 8 & 1 & 6 \\ 3 & 5 & 7 \\ 4 & 9 & 2 \end{pmatrix}$$

The American Museum of Mathematics in New York has a display where players can play this game—on one side, for the adults, you can see only the numbers $1, 2, 3, \ldots, 15$ listed linearly, but on the other side, for the kids, one sees the matrix, and the kids can usually win because of the tic-tac-toe insight.

7.4 Show that a Nim position is balanced if and only if, when you represent the pile heights in binary, the number of one digits in each place is even.

7.5   Consider the modified version of the game Buckets of Fish, where on each turn a player may take a fish from one bucket and then either add or remove as many fish as desired from any of the buckets to the left. What is the winning strategy? Suppose that a player may add or take away at most one fish from each of the earlier buckets. What then is the winning strategy?

7.6   Consider another modified version of the game Buckets of Fish, where on each turn a player may take one, two, or three fish from a given bucket, and then add or remove any number of fish to each of the buckets to the left. What is the winning strategy? And what if players can take up to four, or up to five, and so on?

7.7   Consider a further modified version of the game Buckets of Fish, where on each turn a player may take any positive number of fish from any given bucket and add any number of fish to each of the buckets to the left. What is the winning strategy? [Hint: Would it be a good idea for a player to leave the buckets in such a way that the leftmost two buckets have the same number of fish, and also the next two, and the next two, and so on?]

7.8   Define and prove a sense in which the balancing strategy is the only winning strategy in Nim.

7.9   What is the winning strategy for misère Nim? Prove your answer. [Hint: Prove that a move is a winning move in misère Nim if and only if it is a winning move in Nim, except in the case that it leads to a position with all piles having only one coin.]

7.10   Suppose that we have a Gold Coin game position with pennies at positions $n_0 < n_1 < \cdots < n_k$, and a gold coin at position n. How many moves are in the longest possible legal play?

7.11   Give a complete analysis of the game of Chomp for a 3×2 chocolate bar. Would you rather go first or second? What is the winning play?

7.12   Describe a winning strategy in the game of Chomp for a square chocolate bar $n \times n$, with $n > 1$, and prove that it is winning.

7.13   How large is the game tree for tic-tac-toe? Perhaps it is difficult to say exactly, but what are the best upper and lower bounds you can prove? Does the game have certain symmetries that allow you to understand the game with only part of the game tree?

7.14   How many nodes are there in the game tree of chess on the level after two moves for each player?

7.15   Criticize the play of both players in the tic-tac-toe game illustrated in figure 7.1. Did the winning player play well? On which move did the losing player go wrong?

7.16   Describe a natural version of three-dimensional tic-tac-toe, and prove that the first player has a winning strategy.

Credits

Although forms of the game of Nim have reportedly been known since ancient times, the mathematical analysis of the game is due to Charles L. Bouton (1901). The fundamental theorem of finite games was proved by Zermelo in 1913.

8 Pick's Theorem

Pick's theorem is a mathematical gem, showing that one can compute the area of a polygon formed by vertices in the integer lattice simply by counting the number of vertex points in the interior and on the boundary. How surprising! Let us explore this theorem and its proof, which I find to be an excellent instance of how a difficult problem can be solved by breaking it into easier instances. Allow me to present the theorem as a case study in the advice of George Pólya:

> If you cannot solve a problem, then there is an easier problem that you cannot solve: find that problem. –Pólya (1973)

8.1 Figures in the integer lattice

To begin, consider the *integer lattice*, consisting of the intersection points of the horizontal and vertical integer lines in the plane, the points (a, b), where both a and b are integers. By connecting such points with line segments, we can form polygons in the integer lattice; a *simple polygon* is one formed by a closed-loop sequence of nonintersecting line segments in this way.

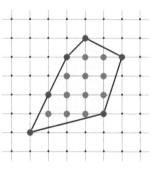

Theorem 63 (Pick's theorem). *The area of a simple polygon formed by vertices in the integer lattice is precisely*

$$A = i + \frac{b}{2} - 1,$$

where i is the number of lattice vertex points in the interior of the polygon and b is the number of lattice vertex points on the boundary.

In the figure above, we have 11 interior points and 6 boundary points, and so the theorem tells us that the area is $11 + 6/2 - 1 = 13$.

How is it possible to prove such a general theorem? We seem faced with a staggering breadth of instances: all simple polygons of any shape and size in the integer lattice. Perhaps we can hope for progress by following Pólya's advice to consider some easier cases?

8.2   Pick's theorem for rectangles

For example, let us consider the case of a rectangle, oriented along the lattice lines. Can we verify Pick's theorem in this easy case?

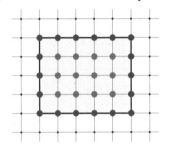

$$i = 12$$
$$b = 18$$
$$A = i + b/2 - 1$$
$$= 12 + 18/2 - 1$$
$$= 20 = 5 \cdot 4$$

As you can see, the theorem is verified for the 5×4 rectangle pictured here, because in this case there are twelve interior vertices and eighteen boundary vertices, and so Pick's formula gives a value of 20, which is indeed the correct area for a 5×4 rectangle.

What about the general case of an $n \times m$ rectangle? Well, we know the correct area should be nm. Let us count the number of interior and boundary vertices. Since the rectangle begins and ends on vertex points, we actually have $n + 1$ and $m + 1$ many vertices on each of the horizontal and vertical sides of the rectangle. But this would be double-counting the corner vertices, which are each on two sides. Therefore, the total number of boundary vertices is

$$b = 2(n + 1) + 2(m + 1) - 4 = 2n + 2m.$$

Similarly, for the interior vertices, there are $n - 1$ rows and $m - 1$ columns, and so the number of interior vertices is

$$i = (n - 1)(m - 1).$$

If we put these together into Pick's formula, we get

$$A = i + \frac{b}{2} - 1$$
$$= (n - 1)(m - 1) + (2n + 2m)/2 - 1$$
$$= nm - n - m + 1 + n + m - 1$$
$$= nm.$$

Since this is indeed the area of an $n \times m$ rectangle, we have therefore proved Pick's theorem in the easy case of a rectangle aligned with the lattice lines. Let us state this as a separate theorem; we have already just given the proof.

Theorem 64 (Pick's theorem, rectangle case). *For any nondegenerate rectangle in the integer lattice, whose sides are aligned with the lattice lines, the area is*

$$A = i + \frac{b}{2} - 1,$$

where i is the number of interior lattice points and b is the number of lattice points on the boundary of the rectangle.

8.3 Pick's theorem for triangles

Let us now try a somewhat harder case. How about a triangle? Well, what kind of triangle? Perhaps we should start with a right triangle, whose legs are oriented with the lattice lines. We'll have to pay attention to whether there are boundary points on the hypotenuse. For the 7×3 triangle pictured here, we have six interior vertices and eleven boundary vertices (none on the hypotenuse), so Pick's formula

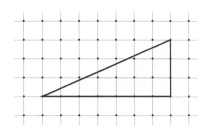

gives $6 + 11/2 - 1 = 10.5$, which is indeed the area of this triangle $\frac{1}{2} \cdot$ base $\cdot$ height $= \frac{1}{2} \cdot 7 \cdot 3 = 10.5$. Let us try to consider the more general case.

Theorem 65 (Pick's theorem, right triangle case). *For any nondegenerate right triangle in the integer lattice, whose legs align with the lattice lines, the area is given by Pick's formula*

$$A = i + \frac{b}{2} - 1,$$

where i is the number of interior vertices and b is the number of vertices on the boundary.

Proof. Consider such a triangle, with legs of length n and m. The hypotenuse of the triangle is the diagonal of a corresponding $n \times m$ rectangle. The area of the triangle is $nm/2$, exactly half of the area of the corresponding rectangle. Let h be the number of boundary points on the hypotenuse, not counting the vertices of the triangle. The triangle has exactly half of the interior points of the rectangle, except for the h points on the hypotenuse itself. In the proof of theorem 64, we calculated the number of interior points of the rectangle as $(n-1)(m-1)$, and so for the triangle, we have $i = [(n-1)(m-1) - h]/2$. For the boundary, the triangle has exactly half of the boundary vertices of the rectangle, plus the h extra points on the hypotenuse, plus one extra, since we have one horizontal side, one vertical side, but we include three corners instead of only two. So for our triangle, we have

$b = n + m + h + 1$. Pick's formula results in

$$i + \frac{b}{2} - 1 = [(n-1)(m-1) - h]/2 + (n + m + h + 1)/2 - 1$$
$$= nm/2 = A.$$

Thus, Pick's formula is correct for such triangles.   □

How about the more difficult case of an arbitrary triangle? Perhaps we might hope to prove the theorem by dividing a general polygon into triangles, thereby reducing to the triangular case.

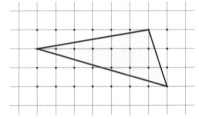

We know how to handle rectangles and right triangles that align with the lattice lines. Can we somehow reduce the case of an arbitrary triangle to these shapes? Alas, even for easy cases such as the triangle above, it does not appear possible.

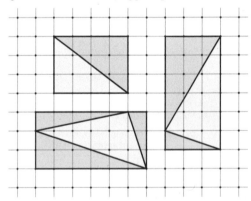

Nevertheless, Pick offers an inspired idea. Namely, instead of reducing to *smaller* right triangles and rectangles, let us instead add right triangles so as to form a larger rectangle. Every triangle in the integer lattice can be completed to a rectangle, oriented along lattice lines, by adding at most three right triangles, also oriented with lattice lines. And crucially, we already know Pick's theorem for the resulting rectangles and the triangles that we had added to form them.

8.4   Amalgamation

Generalizing a bit, the inspired idea is to figure out how Pick's formula is affected when one adjoins one figure to another. In the general case, we want to amalgamate any kind of figures, not just triangles and rectangles. And actually, when we formulate the issue this way, we see that it becomes kind of easy, even though it is a little more abstract.

Key Lemma 66. *Suppose that P and Q are simple polygons in the integer lattice, which do not overlap but which are joined for a connected stretch of one or more edges on their boundaries. Let PQ denote the amalgamated polygon obtained by joining them together.*

1. *If Pick's theorem holds for P and Q separately, then it holds for their amalgamation PQ.*
2. *If Pick's theorem holds for the amalgamation PQ and for one of P or Q, then it holds also for the other.*

In other words, Pick's theorem is preserved by the addition or removal of a simple polygon already satisfying Pick's theorem.

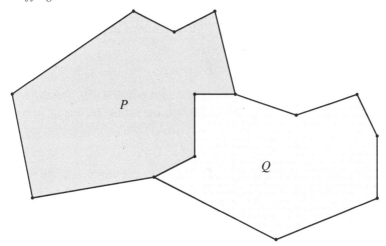

Proof. Let us introduce the notation A_P, A_Q, i_P, i_Q and b_P, b_Q to refer to the area, interior, and boundary lattice-point counts of P and Q, respectively, and similarly with A_{PQ}, i_{PQ}, and b_{PQ} for the amalgamated polygon. For statement (1), therefore, we assume that

$$A_P = i_P + \frac{b_P}{2} - 1$$

and that

$$A_Q = i_Q + \frac{b_Q}{2} - 1.$$

Notice that the area of the amalgamated polygon PQ is simply the sum of the areas of P and Q separately. Let c be the number of boundary lattice points on the common boundary of P and Q, not including the two endpoints of the common boundary. It follows that

$$i_{PQ} = i_P + i_Q + c,$$

since the common boundary points (except for the two endpoints) become interior to the amalgamated region.

Similarly,

$$b_{PQ} = b_P + b_Q - 2c - 2,$$

since the $c + 2$ common boundary points were each counted twice, but only two of them remain on the boundary of the amalgamated region PQ.

Putting all of this together, we may calculate

$$A_{PQ} = A_P + A_Q$$
$$= \left(i_P + \frac{b_P}{2} - 1\right) + \left(i_Q + \frac{b_Q}{2} - 1\right)$$
$$= (i_P + i_Q + c) + \left(\frac{b_P + b_Q - 2c - 2}{2}\right) - 1$$
$$= i_{PQ} + \frac{b_{PQ}}{2} - 1.$$

And this verifies Pick's formula for the amalgamated polygon PQ, thereby establishing statement (1). The reader will prove statement (2), using similar reasoning, in exercise 8.5.

□

One can also prove the lemma with a softer argument and one that, to my way of thinking, provides a deeper explanation of why Pick's formula is preserved by amalgamation. Namely, the boundary points of P and Q count only half in Pick's formula, when computed separately for each figure, because of the $\frac{b}{2}$ term, but in the amalgamated figure, they become interior and therefore count fully in the i_{PQ} term. And the two endpoints of the common boundary were each counted twice (but half each), which exactly balances the extra -1 term from calculating the formulas separately. This explanation can also be seen as what is really going on in the algebraic calculations above.

Let us now use the lemma to implement our idea concerning triangles.

Theorem 67 (Pick's theorem, triangle case). *For any triangle in the integer lattice, the area is given by Pick's formula*

$$A = i + \frac{b}{2} - 1,$$

where i is the number of lattice points interior to the triangle and $b = 3$ is the number of lattice points on the boundary.

Proof. For any such triangle, we may adjoin one, two, or three right triangles so as to form a rectangle, as shown previously. Since Pick's theorem holds for the rectangle and for each of the triangles, it follows from the key lemma 66 that Pick's theorem holds after removing these right triangles one at a time, and so in this way, Pick's theorem holds for the original triangle.   □

8.5 Triangulations

A final ingredient we shall need for a full proof of Pick's theorem in the general case is the fact that every simple polygon in the integer lattice can be triangulated by triangles in the integer lattice.

Lemma 68. *Every polygon admits of a triangulation. That is, the polygon can be partitioned into triangles using the vertices of the polygon, in such a way that the triangles overlap only on their edges and the union of the triangles is the original polygon.*

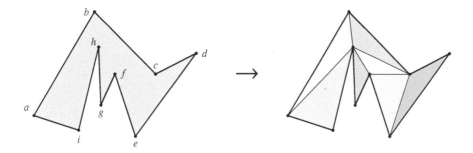

Proof. Since the sum of the interior angles of an n-gon is $180(n-2)°$, it cannot occur that all n vertices have interior angles exceeding $180°$, and so there must be a *convex* vertex, a vertex whose interior angle is less than $180°$. Fix such a convex vertex, such as vertex d in the figure above, and consider the triangle formed by the adjacent sides.

If this triangle is totally contained in the interior of the polygon, as it is in the figure with $\triangle cde$, then we may fix this triangle as part of our triangulation, and by clipping it off, we reduce to a polygon with one fewer side, which by induction has a triangulation, and so in this case we are done.

It remains to consider the case of a convex vertex, such as vertex b in the figure, such that the triangle $\triangle abc$ formed by the adjacent sides is not contained in the interior of the polygon. It follows by convexity that there is at least one vertex of the polygon in the interior of this triangle, as with vertices h and f in the figure. One of these vertices interior to $\triangle abc$, let us call it h, forms the smallest possible angle $\angle hab$. It follows that $\triangle abh$ is contained in the interior of the polygon, because otherwise there would be a vertex forming a smaller angle. It follows that the edge bh lies in the interior of the polygon, and since b

and *h* are not adjacent, this edge cuts the polygon into two smaller simple polygons, one starting at *b* and going to *a* and around to *h*, and the other starting at *b* and going to *c* and around to *h* in the other direction. By induction, each of those smaller polygons has a triangulation, and the union of these is a triangulation of the original polygon.   □

8.6   Proof of Pick's theorem, general case

Let us now prove Pick's theorem in the fully general case.

Proof of theorem 63 (Pick's theorem). Suppose that *P* is a simple polygon in the integer lattice, whose vertices are lattice points. By lemma 68, there is a triangulation of *P* into triangles. By theorem 67, we know that Pick's theorem holds for each of these triangles. By the key lemma 66, Pick's theorem continues to hold as we amalgamate these triangles together. So Pick's theorem holds for the polygon *P*.   □

One can view the proof of Pick's theorem as a proof by induction on the size of the triangulation, that is, by induction on the number of triangles needed in the triangulation. If the theorem is true for all triangulations involving *n* triangles, then the key lemma allows us to add one more.

In exercise 8.10, the reader will consider the possibility of a three-dimensional analogue of Pick's theorem, which might express the volume of a polyhedron in the three-dimensional integer lattice in terms of the number of interior and boundary vertices.

Mathematical Habits

Consider an easier problem. When faced with a difficult problem, consider an easier case, whose solution may form the basis of a general solution. Make a simplifying assumption, consider a special case, or prove a weaker conclusion. The Hungarian mathematician George Pólya, author of a famous book on mathematical problem-solving Pólya (1973), gives the advice, "If you cannot solve a problem, then there is an easier problem you cannot solve: find that problem." You will make progress by cracking that easier problem first.

Build on your results. Having proved a theorem, freely rely on it in new arguments, without need to prove it again. Prove additional theorems and corollaries that build on what you have established. The strongest mathematicians I know seem immediately to absorb new mathematical knowledge and begin drawing out consequences straightaway.

Exercises

8.1   Give a direct proof, without using the results proved in this chapter, that Pick's theorem holds for triangles of height 1, whose base is parallel to one of the axes.

8.2   Give a direct proof of Pick's theorem for squares in the integer lattice, whose diagonals (rather than sides) are oriented with the axes.

8.3   Can we omit the "simple" qualifier in Pick's theorem? [Hint: For example, does Pick's theorem hold for degenerate triangles (with zero area) or for such figures as the bow tie or the annulus, shown below?]

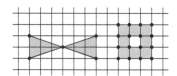

8.4   Prove that every simple polygon in the integer lattice can be triangulated by triangles whose sides contain no lattice points. [Hint: Using lemma 68, it suffices to consider triangles. Argue by induction on the total number of interior and boundary points. If you have got a triangle with a boundary vertex on a side, then make two smaller triangles and use the induction hypothesis.]

8.5   Prove statement (2) of the key lemma 66.

8.6   Does the proof of the key lemma (lemma 66) work in the case that the common part of P and Q is a single vertex, rather than a side? Prove your answer.

8.7   Is the key lemma true when the common boundary of P and Q is not connected? For example, perhaps the figures are joined by two or more connected common boundaries simultaneously. [Hint: Consider exercise 8.3.]

8.8   Can you construct a triangle formed by vertices of the integer lattice with area exactly 3.25?

8.9   Prove that every rectangle formed by vertices in the integer lattice (not necessarily oriented with the axes) has an area that is an integer.

8.10   In three dimensions, consider the tetrahedron with vertices at $(0,0,0)$, $(1,0,0)$, $(0,1,0)$, and $(1,1,n)$ for arbitrary integer n. Draw the tetrahedron, and using the formula for the volume of a cone over any planar figure,

$$\text{volume} = \frac{1}{3}(\text{area of base}) \cdot \text{height},$$

show that the volume is $n/6$. Prove that there are no interior lattice points and no lattice points on the boundary, other than those four. Conclude that there is no three-dimensional analogue of Pick's theorem expressing the volume of a polyhedron solely in terms of the number of interior and boundary vertices.

Credits

Pick's theorem is due to Georg Alexander Pick (1899). The result of exercise 8.10 is due to John Reeve in 1957. I am grateful to Kevin O'Bryant for suggesting the topic of Pick's theorem. The Pólya quotation is attributed to him by John Conway in the foreword of Pólya (1973), but it seems that the quotation may be Conway's summary of Pólya's advice. Pólya is often also credited with saying, "If you cannot solve a problem, then there is an easier problem you *can* solve: find that problem." It seems that neither quotation can be found word for word in Pólya (1973), although Pólya certainly makes the point there in other words. I find Conway's "cannot" formulation erudite, since the "can" formulation suffers from the fact that by trivializing a problem one finds an easier problem that one *can* solve; it is rather more valuable to have an easier but still difficult problem, for perhaps by solving it one will be led eventually to a solution of the original problem.

9 Lattice-Point Polygons

9.1 Regular polygons in the integer lattice

Consider a regular square lattice, the *integer lattice*, the intersection points formed by a grid of uniformly spaced parallel horizontal and vertical lines in the plane. One may easily find vertex points in that lattice that form the vertices of various larger squares. What other types of regular polygons we might find in this lattice?

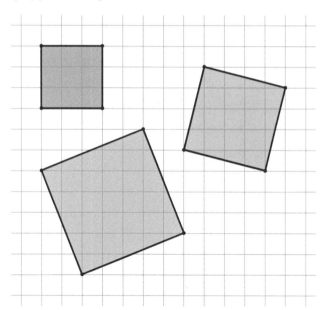

Question 69. Can one find vertex points in the integer lattice that form the vertices of a regular hexagon? Or a regular pentagon? An equilateral triangle? Which regular polygons can be found in the integer lattice?

Go ahead and try! Which regular polygons can you find?

The surprising answer is that one cannot find any regular polygon in this way other than squares.

Theorem 70. *There is no regular pentagon in the integer lattice, and no regular hexagon, no regular heptagon, and so on. Indeed, the only nondegenerate regular polygons to be found using vertices in the square lattice are squares themselves.*

Proof. Suppose that we could find five vertices in the integer lattice that form a regular pentagon. Construct on each side a perpendicular of the same length, as pictured here in brown.

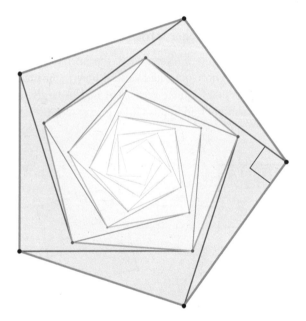

Since the lattice is invariant under rotations by 90° centered at a lattice point, each of those new points is still a lattice point. And by symmetry, they form the vertices of a (slightly smaller) regular pentagon. It follows that there can be no regular pentagon whose side lengths are as small as possible. Therefore, there can be no regular pentagon at all in the lattice, since if there were one, there would have to be one whose side lengths are as small as possible.

We could have argued alternatively like this: there can be no regular pentagons in the lattice, since by iteratively applying the transformation, we would eventually construct a lattice-point pentagon whose size was smaller than the square lattice spacing, which is clearly impossible.

A similar argument works with larger regular polygons. The main point to realize is that for all regular *n*-gons, where *n* > 4, when you construct the perpendicular on one of the sides, the resulting point is strictly inside the original polygon, and this is why the resulting regular *n*-gon is strictly smaller than the original.

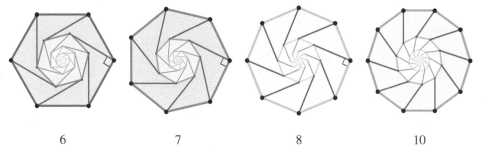

| 6 | 7 | 8 | 10 |

This completes the proof for all *n*-gons for $n > 4$.

The case of an equilateral triangle, however, requires special care. If one attempts the same construction idea as above, building the perpendicular on the edges of a triangle, then the resulting triangles becomes larger and larger, causing the proof method to break down.

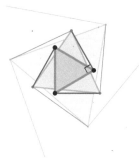

Nevertheless, one can reduce the equilateral triangle case to a hexagon: if you could find an equilateral triangle in the square lattice, then since the lattice is invariant under the rigid translation along any lattice-point line segment, it follows that we could build a regular hexagon. Two such translations along the sides of the triangle suffice to produce all six points on the hexagon.

But we have already showed that we cannot find a regular hexagon in the square lattice, and so we cannot find an equilateral triangle. □

9.2   Hexagonal and triangular lattices

Consider next the lattice arising from the hexagonal tiling of the plane. In this lattice, it is easy to find lattice points that form the vertices of an equilateral triangle or a regular hexagon.

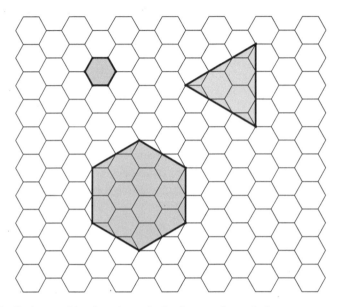

We shall similarly consider the triangular lattice, as shown below.

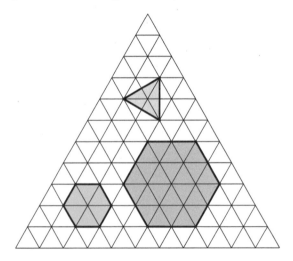

We can easily find equilateral triangles and regular hexagons in the triangular lattice.

In fact, the triangular and hexagonal lattices support exactly the same shapes. One can construct a shape at lattice points in the hexagonal lattice if and only if one can also do so in the triangular lattice.

Theorem 71. *Exactly the same shapes (up to scaling of size) arise as polygons using lattice points in the hexagonal lattice as in the triangular lattice.*

Proof. The main point is that the two lattices refine each other, in the sense that there is a subcollection of vertices in the hexagonal lattice that form a regular triangular lattice, and there is a subcollection of the triangular lattice that form a regular hexagonal lattice, as pictured here.

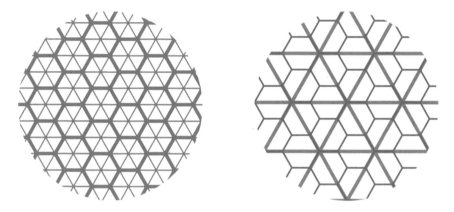

Given a red triangular lattice (at left), we may find lattice points forming the blue overlayed hexagonal lattice. And given a blue hexagonal lattice (at right), we may find lattice points forming the red overlayed triangular lattice. Therefore, any shape that can be formed from vertices in the regular hexagonal lattice can also be formed (up to scaling) from vertices in the triangular lattice and vice versa. □

The question now is which shapes those might be.

Question 72. Which regular polygons can be formed by vertices of the hexagonal or triangular lattices?

The answer is that, beyond the triangles and hexagons, there are no other regular polygons to be found in those lattices.

Theorem 73. *The only nondegenerate regular polygons that arise from vertices in the hexagonal or triangular lattices are the equilateral triangles and regular hexagons.*

This theorem extends the analysis we gave above for the square lattice.

Proof. By theorem 71, it suffices to consider only the triangular lattice. This lattice is invariant by 60° rotation around any vertex point, so the same idea we used in the proof of theorem 70 will work here, except that we shall rotate by 60° instead of 90°.

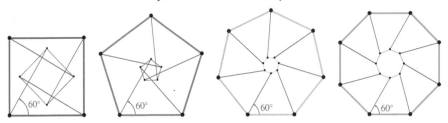

It is easy to see that, except in the case of $n = 3$ and $n = 6$, for the equilateral triangle and regular hexagon, rotating the side of an n-gon by 60° moves the vertex to the interior and produces a strictly smaller regular n-gon. Therefore, as we argued above in the proof of theorem 70, there can be no smallest such regular n-gon in the triangular lattice, except for triangles and hexagons, and consequently no such regular n-gons can arise in this way at all. And therefore by theorem 71, this is also true for the hexagonal lattice.   □

Theorem 74. *For any collection of five or more points in the integer lattice, there are two of them for which the line segment joining them contains another lattice point.*

Proof. Consider any five of the lattice points, and look at the parities of their (n, m) coordinates. There are only four possible parity patterns—(even, even), (even, odd), (odd, even) and (odd, odd)—and so two of the points must have the same parity in each coordinate. Thus, there are two of the five points that differ by a multiple of 2 in each of the coordinate directions. It follows that the midpoint of those two points is a lattice point on the line joining them.   □

In the exercises, the reader will generalize this idea to higher dimensions.

9.3   Generalizing to arbitrary lattices

Let us extend the result to arbitrary lattices. A *lattice* is a discrete collection of points in the plane or in some higher dimension, which is closed under vector sum and differences. In other words, the displacement of any point in the lattice by the vector difference of two lattice points is still a lattice point. For example, the integer lattice has this feature, since the displacement of an integer lattice point (a, b) by an integer amount in each coordinate remains an integer lattice point. The triangular lattice is also a lattice in this sense, but the reader will verify in the exercises that the vertices of the hexagonal lattice, as described in this chapter, do *not* actually form a lattice in this sense. Meanwhile, there are abundant other natural lattices, including rectangular lattices, parallelogram lattices, and cubic lattices, and these are widely studied and important in many applications.

Lemma 75. *Any parallelogram with three points in a lattice also has its fourth point in the lattice.*

Proof. Suppose that $ABCD$ is a parallelogram, with A, B, and D being lattice points. Notice that C is the point resulting from adding vector AD to B, so it is a lattice point. $\square$

Theorem 76. *There are no regular polygons in any lattice, except triangles, squares, and hexagons.*

Proof. Suppose that we have a regular n-gon, whose vertices arise as lattice points in a lattice, where either $n = 5$ or $n \geq 7$. We may assume that we are considering a smallest instance, in the sense of containing the smallest number of lattice points in the interior. For every pair of adjacent sides, build the corresponding parallelogram, as highlighted in the figures below. By lemma 75, each such new vertex is a lattice point.

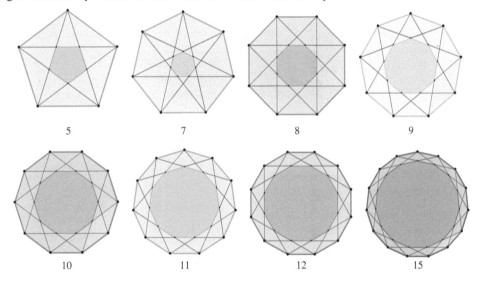

By symmetry, these new lattice points form a strictly smaller regular n-gon, one containing strictly fewer lattice points. This contradicts our assumption that we already had the smallest instance. So there can be no such instance to begin with. $\square$

Mathematical Habits

Eliminate clutter. Edit your proofs to remove irrelevant remarks. Every statement in a proof should be part of the argument.

Iterate consequences. Apply an observation repeatedly. What happens when a process is iterated? Pay attention to what changes and what is preserved to understand the effects of iteration. Sometimes, even a simple idea can gain enormous power under iteration.

Look for a better proof. The well-known chess advice is, if you see a good move, look out for a better one. Similarly, in mathematics, when you have proved a theorem, there might be a better proof lurking nearby. Search for it! Revisit your theorem and proof later; sometimes a different context will enable you to see a streamlined or more elegant argument.

Exercises

9.1   Which regular polygons can be found using vertices of the 2×1 brick tiling? Prove your answer.

9.2   Which sizes of squares arise in the integer lattice? Does $\sqrt{5}$ arise? How about $\sqrt{17}$? Which numbers arise exactly? Can you give a complete if-and-only-if characterization?

9.3   Prove that in the square lattice, any line segment joining two lattice points is the side of a square whose vertices are lattice points.

9.4   Prove or refute the following: In the hexagonal lattice, any line segment joining two lattice points is the side of a hexagon whose vertices are lattice points.

9.5   Prove that the square lattice is not necessarily invariant under the reflection swapping any two lattice points.

9.6   Prove or refute the following: The hexagonal lattice is invariant under the reflection swapping any two lattice points.

9.7   Can some nonsquare regular polygons arise from lattice points in some rectangular lattice, not necessarily square? Exactly which regular polygons can arise in such a lattice?

9.8　Prove that in any rectangular lattice, using a rectangle whose side lengths are commensurable, the only regular polygon to be found using lattice points is a square.

9.9　Using colored chalk on the tiled plaza in the town's market square, a child connects the centers of some of the hexagons like this:

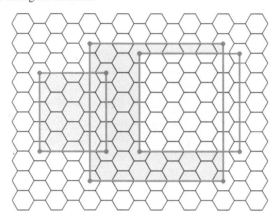

　　Has she made any squares?

9.10　Find the scaling values h and t making the following statement true: Any figure realized in the triangular lattice can be realized in the hexagonal lattice, when scaled by a factor of h; and any figure realized in the hexagonal lattice can be realized in the triangular lattice, when scaled by a factor of t. [Hint: Use the proof of theorem 71.]

9.11　Prove that there are no regular polygons, except squares, in the rational plane $\mathbb{Q} \times \mathbb{Q}$. [Hint: For any particular polygon, find a common denominator for the coordinates that are used.]

9.12　Prove that for any nine points in the three-dimensional integer cubic lattice $\mathbb{Z} \times \mathbb{Z} \times \mathbb{Z}$, there are two such that the line segment joining them contains another lattice point. Can one generalize this to n-dimensions? Is this bound optimal?

9.13　Prove that the ratio of any two parallel line segments joining points in the integer lattice is a rational number. Is this true without the "parallel" qualifier?

9.14　Prove that the vertices of the triangular lattice, as described in the chapter, are a lattice in the sense of section 9.3, but the vertices of the hexagonal lattice are not. And do any three points in the hexagonal lattice lie on a lattice parallelogram? That property was important in the proof of theorem 76.

9.15　Explain what happens precisely with the parallelogram argument proving theorem 76 when it is applied to triangles, squares, or hexagons.

9.16　Which of the five Platonic solids can be realized at lattice points in the three-dimensional integer lattice?

9.17   For each integer $n \geq 3$, let $d(n)$ be the smallest dimension such that there is a regular planar n-gon realized at points of rational coordinates in dimension $d(n)$, if possible. For which values of n is $d(n)$ defined, and what is the complete list of values of the function $d(n)$?

Credits

I learned the main idea of this chapter from a post by Vaughn Climenhaga on MathOverflow, Climenhaga (2017); he had presented the hexagonal case, crediting the idea to György Elekes during an instance of the Conjecture and Proof course in the Budapest Semesters in Mathematics. The proof of theorem 76 grew out of an exchange among myself, John Baez, and Yao Liu on Twitter at https://twitter.com/JDHamkins/status/1154865078493208576. Exercise 9.9 was inspired by math teacher Kate Belin in her sidewalk-math tweet at https://twitter.com/katebelin/status/1161701927496929282. Exercise 9.17 was suggested by David Madore at https://twitter.com/gro_tsen/status/1161660140879306754. The chess adage about looking for a better move appears in annotations of William Wayte (1878).

10 Polygonal Dissection Congruence Theorem

Have you ever found yourself faced with a mathematical statement of sweeping generality, easily verified in numerous cases but which nevertheless remained mysterious? We can check many instances—indeed, we might be able to check any instance we care to think about—and yet, we may lack insight into why the statement should necessarily be true for all cases. Despite what may be abundant evidence in the particular cases, these separate arguments may continue to seem ad hoc in their particularity, not yet unified by a general explanation. One can sometimes seem to know all the specific cases individually, separately, without comprehending the general fact.

10.1   The polygonal dissection congruence theorem

To my way of thinking, the polygonal dissection theorem of Wallace, Bolyai, and Gerwein is such a case. The theorem asserts that if two polygons have the same area, then they are *dissection congruent*, which is to say that we may cut the first polygon into finitely many polygonal pieces that can be rearranged, like a puzzle, to form the second polygon. Amazing!

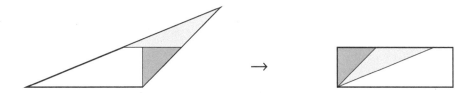

Theorem 77 (Wallace-Bolyai-Gerwein, 1807, 1833). *Any two polygons of the same area are dissection congruent.*

We have abundant instances of the theorem, because whenever we cut a shape into pieces and rearrange them, we have thereby observed another instance of the theorem. The deep mystery of the theorem is that it holds generally, provided only that the same-area requirement is fulfilled.

Note that the same-area sufficient condition is also necessary, because if two polygons are dissection congruent, then they are formed like a puzzle from the same puzzle pieces. The area is the total area of the pieces in each case. So we could have stated the theorem as a biconditional: two polygons are dissection congruent if and only if they have the same area.

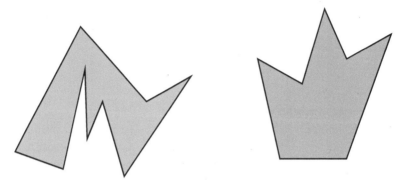

The mystery is the general case. Faced with two equal-area shapes, perhaps very complicated, how shall we even begin to find the common partition? It may not be clear even how to begin.

10.2   Triangles to parallelograms

Let us try. Perhaps we can simplify the problem by reducing from the case of an arbitrary polygonal shape to something more specific. We know by lemma 68, after all, that every polygon admits of a triangulation, a partition of the polygon into disjoint triangles. So let us consider the dissection theorem in the case of triangles. Aha! With any individual triangle, we can make progress.

Lemma 77.1. *Every triangle is dissection congruent to a parallelogram.*

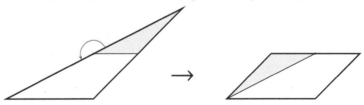

Proof. Given any triangle, consider one of the edges as a base. Cut the triangle at half height, and swing the upper part down as shown, rotating that part by 180°. This forms a parallelogram, because each of the sides was bisected and the half-height horizontal is half the base, so we have a figure with opposite sides congruent, which is a parallelogram.   □

10.3 Parallelograms to rectangles

Lemma 77.2. *Every parallelogram is dissection congruent to a rectangle.*

Proof. Place the parallelogram upon one of its longest sides as a base.

It follows that there is a vertical from that side to its opposite (why?). By slicing on that vertical and swapping the order of the two sides, we form a rectangle, as shown. □

10.4 Rectangles to squares

Lemma 77.3. *Every rectangle is dissection congruent with a square.*

Proof. If we are faced with a long, thin rectangle, we can make it not quite as long and thin by cutting it in half and stacking the two halves, like so:

This will double one side length and halve the other. By iterating this, if necessary, we produce a rectangle with neither side more than twice as long as the other.

Let s be the side length of the target square we are trying to construct from the rectangle by dissection. This is simply the square root of the area of the rectangle.

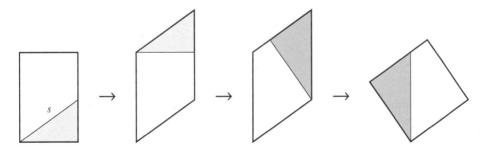

Place the rectangle on its short side at the base, and construct a line of length s from the lower left corner to the side at right, as shown. Slice on that line, cutting off the gold triangle, and place this triangle above to form a parallelogram.

Next, construct the perpendicular at the new lower right vertex, forming the green triangle. This line meets the opposite side, because of our assumption that the rectangle was not more than twice as tall as it is wide (why?). By translating the green triangle, we form a rectangle. Since one side is s and the area is the same as the original rectangle, the other side must also be s, and so this is the desired square.   □

10.5 Combining squares

We shall now seek to combine the squares together.

Lemma 77.4. *Any two squares joined together are dissection congruent with a single larger square.*

Proof. This argument amounts to one of the classical dissection proofs of the Pythagorean theorem.

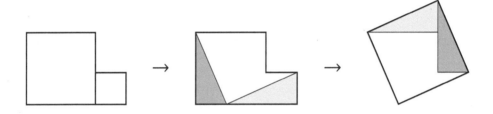

Namely, place the two squares adjacent to each other as shown, and slice off two right triangles whose legs are congruent with the sides of the squares. The triangles can be reassembled as shown to form a larger square.   □

In the arguments above, we were actually making a subtle use of the following fact, which I should like now to highlight.

Observation 78. *The dissection congruence relation is transitive: if P is dissection congruent to Q and Q is dissection congruent to R, then P is dissection congruent to R.*

Proof. We may cut P into finitely many pieces and rearrange them in order to form polygon Q, and we may cut Q into pieces so as to form R. The cuts we make in Q induce cuts on the individual pieces that we had already made when forming Q out of P. It is as though the individual pieces we had made in P are cut again to form smaller pieces. These subpieces can be seen to provide a partition of P that can be used to form R.   □

10.6 Full proof of the dissection congruence theorem

Let us now combine all of the ideas we discussed above and assemble them into a full proof of the dissection congruence theorem.

Proof of theorem 77. Consider any polygon in the plane. By lemma 68, this polygon has a triangulation, a partition of it into triangles. Each triangle in such a triangulation can itself be cut into pieces and rearranged by lemma 77.1 to form a parallelogram, which by lemma 77.2 can be cut and formed into a rectangle, which by lemma 77.3 can be cut and assembled into a square. These squares can be systematically combined by lemma 77.4 to form larger and larger squares until there is just one large square. In this way, every polygon is dissection congruent ultimately with a square of the same area. In particular, any two equal-area polygons will become dissection congruent to the same-size square, and hence also dissection congruent to each other, as desired. □

10.7 Scissors congruence

So far, we have cut only on straight lines. But suppose we generalize this, allowing ourselves to cut more general kinds of pieces, like a jigsaw puzzle.

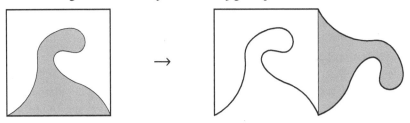

Namely, we define that two figures in the plane are *scissors congruent* when with scissors we can cut the first figure into finitely many pieces, which can be reassembled so as to form the second figure; each cut should follow a simple curve with finite length (a *rectifiable* curve). Which figures are scissors congruent? Perhaps all that is needed is to satisfy the equal-area requirement? How about the following very simple case:

Question 79. Are a square and disk of the same area scissors congruent?

Interlude...

Theorem 80 (Dubins-Hirsch-Karush, 1964). *A square and disk are never scissors congruent.*

Proof. Suppose that we cut a square into finitely many pieces with scissors. If we are to rearrange them into a disk, then some of the cuts must be made by following a circular arc with the scissors, in order to form the outer boundary of the circle. But for every such cut, the pieces on one side would have convex circular-arc boundaries, while the pieces on the other side would have concave circular-arc boundaries. Indeed, because of this observation about how they are made, the total length of the convex circular boundaries of the pieces cut from the square must exactly balance the total length of the concave circular arc boundary segments. And this will remain true after we rearrange the pieces. If we could form a disk with these pieces, then some of the convex circular arc boundaries must be used to form the outer boundary of the disk. And pieces joining in the interior of the disk would cancel as before. In order to form a disk, therefore, the convex circular arc boundary portion of the pieces must strictly exceed the concave parts. Therefore, there is no way to cut a square into pieces with scissors and rearrange them to form a disk.   □

Essentially the same idea can be used to establish many other similar results.

Theorem 81. *No disk is scissors congruent to two smaller disks.*

Proof. Let R be the radius of the original large disk, and let r_1 and r_2 be the radii of the two smaller disks. Since the boundary of the original disk cannot be used on the boundary of the smaller disks, it follows that the scissors must have cut along circular arcs of radii r_1 and r_2 in the interior of the original disk. But in this case, the pieces would be balanced between convex and concave on the r_1-circular arc boundary lengths of the pieces. In the two-disk arrangement, they cannot be balanced in this way, since the outer boundary of the r_1-disk will exhibit an excess of convex pieces in order to make its outer edge.   □

We can generalize these arguments to the following.

Theorem 82. *If two figures, bounded by rectifiable simple closed curves, are scissors congruent, then for every positive radius r, the difference between the convex and concave lengths of radius-r circular arcs on their boundaries must be the same.*

In other words, the radius r convex- minus concave-boundary length difference is an *invariant* of the scissors congruence relation, since any instance of the congruence will have the same difference before and after.

Proof. As in the earlier arguments, cuts made in the interior of P or Q will add convex and concave arcs equally, and they will therefore preserve the convex minus concave difference. So whatever difference exists already on the boundary of P must be preserved to the boundary of Q.   □

Any invariant of an equivalence relation can be used to prove instances of inequivalency, such as we have just proved in several cases for scissors congruence. In this way, we may view theorems 80 and 81 as corollaries of theorem 82. And one may prove many similar theorems using other shapes, by identifying suitable invariants.

Let us generalize the idea of dissection congruence to higher dimensions. What happens for three-dimensional solids? Imagine slicing a three-dimensional figure as though with a knife, cutting a cube or tetrahedron on various planes, and then reassembling the pieces into a different shape. This kind of dissection and reassembly will preserve volume, of course, but can we expect a sweeping general result as with two-dimensional dissection congruence theorem? A direct three-dimensional analogue of the Wallace-Bolyai-Gerwein result (theorem 77) would lead us to expect that any two equal-volume polytopes in three dimensions should be dissection congruent. But is this true?

Question 83 (Hilbert). Must polytopes of the same volume be dissection congruent?

This question was third on the list of questions famously asked by David Hilbert at the dawn of the twentieth century, questions that in many cases guided decades of twentieth-century mathematical research. Some of Hilbert's questions remain open to this day. This particular question, however, was answered by Max Dehn in 1902, when he proved that a cube and a regular tetrahedron are not dissection congruent. You cannot slice a tetrahedron into finitely many polytope pieces, such that you can reassemble them into a cube. In order to prove this, Dehn introduced what is now called the *Dehn invariant*, a certain quantity one can calculate for any polytope, which is preserved under dissection congruence, he proved, but which has different values for the cube and the regular tetrahedron. Thus, the polygonal dissection theorem does not directly generalize to three-dimensional polytopes.

Mathematical Habits

Play with your ideas. Approach a new mathematical topic with a spirit of fun. Play around with the ideas to see how things interact. Modify the hypotheses or the goal, or change up the example cases, even in absurd or silly ways, to deepen your understanding of the original idea.

Identify and use invariants. When investigating a process or operation, find something that remains unchanged by it, an *invariant* quantity, which remains the same before and after. Use the invariant to prove instances of nontransformation, where one mathematical structure or situation cannot be transformed by that operation to another, since they disagree on this invariant. By this means, one can often achieve a classification of the phenomenon by that invariant.

> **Become familiar with a rich collection of examples.** With any mathematical topic, explore the range of examples illustrating it. Explore diverse cases, including very easy cases as well as complex ones, and learn how the fundamental ideas play out with them. Know them like friends and family.

Exercises

10.1 Explain what might go wrong in the proof of lemma 77.2 if we do not place the parallelogram on one of its longer sides.

10.2 Prove the claim in the proof of lemma 77.3 that every rectangle is dissection congruent to a rectangle with no side more than twice as long as the other.

10.3 What is the minimum number of pieces needed to exhibit a dissection congruence between a square and a 45-45-90 triangle?

10.4 What is the minimum number of pieces needed to exhibit a dissection congruence between a square and a 2×1 rectangle?

10.5 What is the minimum number of pieces needed to exhibit a dissection congruence between a square and two smaller squares?

10.6 Draw careful triangles of various shapes on a piece of stiff paper. Using scissors, carry out the dissection congruence of them each with a square of the same area.

10.7 Using an ordinary piece of letter-size paper and scissors, carry out the algorithm of the text to make it dissection congruent with a square.

10.8 True or false: Dissection congruence of polygons in the plane preserves perimeter as well as area.

10.9 Use the figure of lemma 77.4 to give a dissection proof of the Pythagorean theorem.

10.10 There are several dissection congruence notions, depending on whether we allow arbitrary rigid motions with the constituent pieces, or just translation, or just translation and rotation but not reflection. Does this matter? Would the polygonal dissection congruence theorem still hold with these stricter notions of congruence? Prove that we do not ever need to reflect in the dissection congruence theorem—we need not ever flip the puzzle pieces over.

10.11 Prove that no disk is scissors congruent to a noncircular ellipse.

10.12 Prove that two ellipses are scissors congruent if and only if they are congruent.

10.13 Prove that a sphere and a cylinder of the same volume are not dissection congruent (allow cutting along any plane).

10.14 Suppose you form two shapes in the plane using straight lines and arcs from a circle of radius r. Prove that if the two shapes have the same area and used the same fragment of the circle, then they are scissors congruent.

10.15 Consider a generalized scissors congruence concept, allowing infinitely many pieces, provided that the total length of all cuts is finite. Does the analogue of theorem 80 still hold?

Credits

When writing this chapter, I was inspired by a conference presentation of Andrew Marks in which he described his solution with Spencer Unger of Tarski's circle-squaring problem, a considerably more advanced result than what appears here, although Marks drew upon the Wallace-Bolyai-Gerwein theorem and scissors congruence results as motivation (see slides at http://www.math.ucla.edu/~marks/talks/circle_squaring_talk.pdf). See also the construction animation tool of Dima Smirnov and Zivvy Epsteinat at https://dmsm.github.io/scissors-congruence, which uses a slightly different method than here for polygonal dissection congruence.

11 Functions and Relations

11.1 Relations

Mathematics and life generally are full of relations. Consider the sibling relation S on the set of all people, saying that person a stands in the relation S to person b, written

$$a \, S \, b,$$

if a is a sibling of b. We are using infix notation here, placing the relation symbol S between the arguments a and b, just as one would, for example, when writing $x < y$ or $A \subseteq B$. Consider the *child-of* relation $a \, C \, b$, which holds whenever a is a child of b, or the *having-once-cooked-a-meal-for* relation $a \, M \, b$, which holds whenever a once cooked a meal for b. More mathematical relations would include the *less-than* relation $<$ on the set of integers, or the *less-than-or-equal-to* relation $\leq$ on the set of real numbers, or the *subset* relation $\subseteq$ on the collection of sets of natural numbers.

What is a relation, really? Once the domain of discourse is set, a relation is determined by its extension, the information about when it holds and when it does not. Let us therefore define that a *binary relation* on a set A is simply a set of ordered pairs from that set. For any two elements a and b in the set, we write $a \, R \, b$ to indicate that the relation R holds of this pair (a, b) or, in other words, that (a, b) is in R. As a set of pairs, the relation thus consists of exactly the instances for which it holds. Some terminology will enable us to express and refer to some key features a relation might exhibit.

Definition 84. Suppose that R is a binary relation on a set A.

1. The relation R is *reflexive* on A if $a \, R \, a$ for all $a \in A$.
2. The relation R is *symmetric* on A if, whenever $a \, R \, b$, then also $b \, R \, a$.
3. The relation R is *transitive* on A if, whenever $a \, R \, b$ and $b \, R \, c$, then $a \, R \, c$.

Consider, for example, the *divides* relation on the natural numbers, written $a \mid b$, which holds whenever the number a divides the number b, meaning that b is an integer multiple of a. In other words, $a \mid b$ if and only if there is an integer k for which $ak = b$. But kindly be on guard against some common naive misunderstandings of this terminology and notation. Namely, first, the divides relation $a \mid b$ is not at all the same thing as the fractional expression b/a. The expression b/a, after all, represents a number, the numerical result of dividing the number b by the number a, and similarly with a/b; but an instance of the divides relation $a \mid b$, in contrast, is not a number at all—it is a *statement*, a statement describing how a and b are related. The divides relation $a \mid b$ is true exactly when a divides an integral number of times into b. Second, please take care with the order of the numbers in these expressions, because when we write $a \mid b$ for distinct positive integers, for example, then a is the factor and b is the multiple; this is the divides relation, rather than the divisible-by relation.

Theorem 85. *The divides relation $a \mid b$ on the set of natural numbers is reflexive and transitive but not symmetric.*

Proof. (Reflexive) The divides relation is reflexive since every number divides itself: $a \mid a$, since $a \cdot 1 = a$.

(Transitive) To see that the divides relation is transitive, suppose that $a \mid b$ and that $b \mid c$. Since $a \mid b$, there is a number k such that $ak = b$. Since $b \mid c$, there is a number r such that $br = c$. Putting these two facts together, we see that $a(kr) = (ak)r = br = c$, and so we see that $a \mid c$, as desired.

(Not symmetric) The divides relation is not symmetric, since $3 \mid 6$ but it is not the case that $6 \mid 3$. $\square$

Similarly, the reader can easily verify that the usual less-than relation $<$ on the integers is nonreflexive, nonsymmetric, but transitive. The less-than-or-equal-to relation $\leq$, in contrast, is reflexive and transitive but not symmetric.

11.2   Equivalence relations

It is difficult to overstate the importance and ubiquity of the equivalence relation concept in mathematics. This notion arises in nearly every area of pure mathematics and should be seen as a general conceptual tool.

Definition 86. An *equivalence relation* is a relation that is reflexive, symmetric, and transitive.

Equivalence relations generalize the principle features of one of the most important relations there is, the relation of identity $a = b$, which is, of course, reflexive, symmetric, and transitive. In short,

> *equality* is an equivalence relation.

Thus, equivalence relations are a kind of identity relation; we use them when we want to consider softer concepts of identity, involving only the relevant aspects of individuals while ignoring irrelevant differences that would cause two objects to be strictly nonidentical, even when they are identical enough in all the ways that matter for a particular purpose. For example, if you are assembling a machine, you might consider all the type 6 hexhead bolts as equivalent for the purpose of assembling the machine; although the bolts are not identical as objects in the physical universe, they are nevertheless identical enough for the purpose of assembling the machine. Because equivalence relations thus express a concept of identity, mathematicians often use equality-like symbols, such as $\approx$, $\cong$, $\equiv$, and $\sim$, to represent the equivalence relations they are considering, and these symbols are often defined and redefined with different meanings in different situations.

Mathematical examples abound. In the classical geometry of Euclid, similarity of triangles is an equivalence relation. Congruency of geometric shapes is an equivalence relation. The equality relation on rational numbers $\frac{1}{2} = \frac{3}{6}$ amounts to an equivalence relation on the fractional expressions representing those numbers, as the reader will verify in exercise 11.1. In number theory, the relation of having the same parity (even/odd) is an equivalence relation on the integers. In contemporary mathematics, an enormous number of mathematical constructions proceed by defining a certain equivalence relation on a set and then proceeding with that relation as a de facto identity relation.

Consider, for example, the relation of *congruence-modulo*-5 on the integers, commonly written as $a \equiv_5 b$, which holds when a and b have the same remainder when divided by 5, in the sense of lemma 11.

Theorem 87. *The relation of congruence-modulo-5 is reflexive, symmetric, and transitive. Thus, it is an equivalence relation.*

Proof. Every number has the same remainder as itself, when dividing by 5, and consequently, congruence-modulo-5 is a reflexive relation $a \equiv_5 a$. For symmetry, if $a \equiv_5 b$, then a and b have the same remainder when dividing by 5. It follows, of course, that b and a have the same remainder, and so $b \equiv_5 a$. So the relation is symmetric. For transitivity, assume that $a \equiv_5 b$ and $b \equiv_5 c$ modulo 5. So a and b have the same remainder, and b and c have the same remainder, when dividing by 5. So all three numbers have the same remainder, and so in particular, a and c have the same remainder. So $a \equiv_5 c$, verifying this instance of transitivity. □

It often happens in mathematics that one has an equivalence relation $\sim$ on a set A, and one wants to define an operation on the equivalence classes, but one does so by making reference to a particular element of the equivalence class, the *representative* of the class. For example, perhaps one defines $f([a]) = g(a)$, meaning that the function f is meant to take an equivalence class $[a]$ as input, but the value of the function on $[a]$ is defined by applying g to a, which is only one element of $[a]$. This process can be subtly dangerous, because one has succeeded in defining the operation on the equivalence classes, only when g is *well defined* with respect to $\sim$, that is, when $a \sim b \implies g(a) = g(b)$. Let us illustrate with an example.

Theorem 88. *Addition and multiplication of integers are well-defined modulo 5.*

Proof. We want to prove that $x+y$ and $x \cdot y$ are well defined with respect to the equivalence relation $\equiv_5$ of congruence modulo 5, mentioned above. That is, we want to prove that equivalent inputs get mapped to equivalent outputs. So, suppose that $x \equiv_5 x'$ and $y \equiv_5 y'$ are our inputs. It follows that $x = x' + 5k$ and $y = y' + 5r$ for some integers k and r. From this, it follows that $x+y = (x' + 5k) + (y' + 5r) = x' + y' + 5(k + r)$, and so $x+y \equiv_5 x' + y'$. Similarly, $xy = (x' + 5k)(y' + 5r) = x'y' + 5ky' + 5rx' + 25kr = x'y' + 5(ky' + rx' + 5kr)$, and so $xy = x'y'$. The outputs are equivalent. So both addition and multiplication are well defined with respect to congruence modulo 5. $\square$

In this case, we say that the equivalence relation $\equiv_5$ is a *congruence* with respect to addition + and multiplication $\cdot$ on the integers.

Consider the three defining properties of an equivalence relation—reflexivity, symmetry, and transitivity. Are they redundant? Do we really need to check all three? Perhaps one of these properties is a consequence of the other two?

Question 89. Are the properties of a relation being reflexive, symmetric, and transitive redundant?

The answer is no, they are not redundant. But how could we prove such a thing? What we need to do is to exhibit particular relations having exactly two of the properties, but not the third, for each of the combinations.

Theorem 90. *In a binary relation, no two of the properties of reflexivity, symmetry, and transitivity imply the third.*

1. *There is a binary relation that is reflexive and symmetric but not transitive.*
2. *There is a binary relation that is reflexive and transitive but not symmetric.*
3. *There is a binary relation that is symmetric and transitive but not reflexive.*

Proof. For statement (1), consider the *differing-by-at-most-one* relation on the integers, where $a \mathrel{R} b$ just in case a and b differ by at most one, or in other words, $|a - b| \leq 1$. This relation is reflexive, since every number differs from itself by zero, which is at most one. This relation is also symmetric, since if a and b differ by at most one, then b and a differ by at most one. But it is not transitive, since $1 \mathrel{R} 2$ and $2 \mathrel{R} 3$, but 1 is not R-related to 3. So this relation fulfills the desired requirements. Meanwhile, the usual $\leq$ relation on the integers is reflexive and transitive but not symmetric, thereby establishing statement (2). For statement (3), consider the *empty* relation on a set, the relation that never holds. This is vacuously symmetric and transitive. But the empty relation on a nonempty set is not reflexive, since no element of the set is related to itself. It follows that no two of the properties of reflexivity, symmetry, and transitivity can imply the third. $\square$

But hang on. Elsewhere in this book, I have emphasized the mathematical habit of mind that one should distinguish sharply between examples and proof. And surely it is a common mistake amongst beginners to confuse example with proof. Yet, the proof we have just given of theorem 90 appears to have the character of giving examples rather than giving a general argument. Have we made the beginner's mistake? How can we have proved a theorem with an example? The answer is that, indeed, one cannot prove a universal statement by providing an example, and most theorems in mathematics are universal assertions. Nevertheless, theorem 90 is not a universal statement but an existential statement: it is stating that certain kinds of binary relations exist. Such statements can of course be proved simply by exhibiting a successful instance.

11.3 Equivalence classes and partitions

For each object a in the domain of an equivalence relation $\sim$, the *equivalence class* of a, denoted by $[a]$ or by $[a]_\sim$ when one wants especially to clarify which equivalence relation is being used, is the set of all objects equivalent to a:

$$[a] = \{\, b \mid a \sim b \,\}.$$

If $\sim$ is an equivalence relation, then the set of equivalence classes form what is called a partition of the domain.

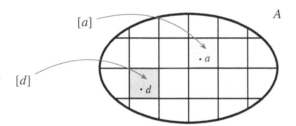

Definition 91. A *partition* of a set A is a family of nonempty pairwise-disjoint subsets of A, whose union is all of A.

To be pairwise disjoint means that if X and Y are distinct sets in the family, then $X \cap Y = \emptyset$. Thus, a partition is a dividing up of the set A into distinct nonoverlapping regions, each containing part of A.

Theorem 92. *Suppose that $\sim$ is an equivalence relation on a set A.*

1. *Every equivalence class $[a]$ is nonempty.*
2. *If equivalence classes $[a]$ and $[b]$ have elements in common, then they are equal.*
3. *The union of all the equivalence classes is A.*

Thus, the set of equivalence classes form a partition of A.

Proof. (1) Since $a \sim a$ by reflexivity, it follows that $a \in [a]$, and so $[a]$ is definitely nonempty.

(2) Suppose that x is a common element of $[a]$ and $[b]$. So $a \sim x$ and also $b \sim x$. To show that $[a] = [b]$, we shall show that $[a] \subseteq [b]$ and conversely. To show this, suppose that $c \in [a]$. So $c \sim a$ and, consequently, $c \sim x$ by transitivity. But since $x \sim b$, we conclude that $c \sim b$ by transitivity again. So $c \in [b]$. Thus, we have proved that $[a] \subseteq [b]$. An essentially identical argument establishes that $[b] \subseteq [a]$, and so they are equal: $[a] = [b]$.

(3) Since $a \in [a]$, the union of all the equivalence classes $\bigcup_{a \in A} [a]$ includes every element of A, and so it is equal to A. $\square$

Let us now consider the converse implication.

Theorem 93. *Every partition on a set A is the set of equivalence classes for some equivalence relation.*

Proof. Suppose that $\mathcal{F}$ is a partition of A. So $\mathcal{F}$ is a family of nonempty subsets of A, which are pairwise disjoint and which union up to A. Let us define a binary relation $\sim$ on A by saying that $a \sim b$ for $a, b \in A$ if and only if there is some $X \in \mathcal{F}$ with $a, b \in X$. In other words, two points are $\sim$-equivalent if they belong in common to a piece of the partition.

Let us argue in detail that $\sim$ is an equivalence relation. First, I claim that $\sim$ is reflexive. To see this, consider an arbitrary element $a \in A$. Since the union of the sets in $\mathcal{F}$ is all of A, there must be some $X \in \mathcal{F}$ with $a \in X$. In this case, a is in the same piece of $\mathcal{F}$ as itself, and so $a \sim a$. So the relation is reflexive. Symmetry is easy, since if $a \sim b$, then there is some $X \in \mathcal{F}$ with $a, b \in X$, and the same X will witness that $b \sim a$. Finally, for transitivity, assume that $a \sim b$ and $b \sim c$. From $a \sim b$ we know that $a, b \in X$ for some $X \in \mathcal{F}$. From $b \sim c$ we know that $b, c \in Y$ for some $Y \in \mathcal{F}$. Since b is in both X and Y, it follows that X and Y are not disjoint, and so we must have $X = Y$ since the sets in $\mathcal{F}$ are pairwise disjoint. Thus, all three points are in the same piece of the partition $a, b, c \in X = Y$. In particular, we know that $a \sim c$, and so the relation is transitive.

Finally, we note that the partition $\mathcal{F}$ is exactly the collection of equivalence classes of $\sim$. The reason is that if $a \in X$ for some $X \in \mathcal{F}$, then every $b \in X$ will be $\sim$-equivalent to a, and only these, and so $X = [a]$. Thus, the original partition $\mathcal{F}$ is exactly the collection of equivalence classes arising from $\sim$. $\square$

Define that one equivalence relation E *refines* another equivalence relation F on the same domain A if $a \, E \, b$ implies $a \, F \, b$ for all $a, b \in A$. For example, the relation of congruency of triangles in the Euclidean plane refines the relation of similarity of triangles, since congruent triangles are necessarily similar. For another example, the relation of congruency modulo 10 in the integers refines the relation of congruency modulo 5, because if two numbers have the same remainder when dividing by 10, then they will also have the same remainder when dividing by 5.

A binary relation R on a set A is *irreflexive* if there is no $a \in A$ with $a \, R \, a$. For example, the child-of relation on the set of all people is irreflexive, since no person is a child of themselves. A relation R is *asymmetric* if, whenever $a \, R \, b$, then $b \, R \, a$ definitely fails. For example, the usual strictly-less-than relation $<$ on the integers is asymmetric, since if $a < b$, then definitely not also $b < a$. A relation is *antisymmetric*, in contrast, if whenever $a \, R \, b$ and $b \, R \, a$, then $a = b$. For example, the less-than-or-equal-to relation $\leq$ on the integers is antisymmetric, since if $a \leq b$ and $b \leq a$, then indeed $a = b$. Finally, we say that a relation is *nonreflexive* if it is not reflexive; it is *nonsymmetric* if it is not symmetric; and it is *nontransitive* if it is not transitive.

11.4   Closures of a relation

The *reflexive closure* of a relation R on a given domain is the smallest reflexive relation $\bar{R}$ on that domain containing R. This is easy to construct, simply by adding in the relational instances $a \, \bar{R} \, a$.

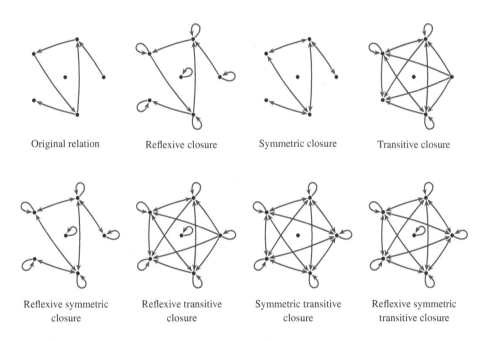

| Original relation | Reflexive closure | Symmetric closure | Transitive closure |

| Reflexive symmetric closure | Reflexive transitive closure | Symmetric transitive closure | Reflexive symmetric transitive closure |

Similarly, the *symmetric closure* of a relation is the smallest symmetric relation containing the given relation, and the *transitive closure* of a relation is the smallest transitive relation containing it. Combining things, the *reflexive-transitive closure* of a relation is the smallest reflexive and transitive relation containing it.

One way to think about the *reflexive-transitive closure* of a relation R on a given domain is that it is the "reachability" relation: $a \, \bar{R} \, b$ just in case there is a *path* from a to b through R, a sequence $a_0, \ldots, a_n$, starting with $a_0 = a$ and ending with $a_n = b$, such that at each step we follow the relation R, so that $a_i \, R \, a_{i+1}$. This is both reflexive and transitive.

Is it possible that different relations on a set have the same reflexive-transitive closure? Yes, of course. If a relation R is not reflexive and transitive, then R is not equal to its reflexive-transitive closure, but closing twice does not add anything more, and so there are two different relations with the same reflexive transitive closure.

11.5   Functions

Let us conclude this chapter with a discussion of functions. What does it mean to have a function f from A to B? Perhaps one is used to considering the *graph* of a function, which in this case would be a certain subset of $A \times B$, the set of pairs (a, b) for which $b = f(a)$. Indeed, all the essential information about the function f is contained in this graph. Because of this, a function can be seen as a special type of binary relation. So let us define that f is a *function* from A to B, written $f : A \to B$, if f is a set of pairs $f \subseteq A \times B$ exhibiting the function property: for every $a \in A$, there is a unique $b \in B$ with $(a, b) \in f$. This object b is denoted by $f(a)$ and is called the *value* of the function at a. If you think about it, the function property corresponds exactly to what is sometimes called the *vertical line test* in precalculus, where every vertical line intersects the graph of the function in at most one point; the vertical line test is exactly testing for the uniqueness of the object b.

We refer to A as the *domain* of the function, and some mathematicians like to count the set B as an essential part of the function, called the *target* or the *codomain*. The *range* of the function, in contrast, is the set of all b that arise as $f(a)$ for some a in the domain. This may or may not be all of the target set B. For example, a constant function $f : \mathbb{R} \to \mathbb{R}$ has a range with only one element in it.

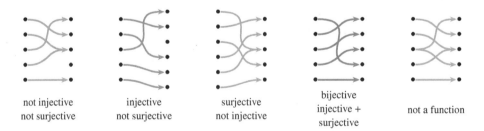

not injective	injective	surjective	bijective	
not surjective	not surjective	not injective	injective +	not a function
			surjective	

A function $f : A \to B$ is said to be *onto* or *surjective* if the range of f is all of B. The function $f : A \to B$ is *one-to-one* or *injective* if, whenever $a \neq a'$, then $f(a) \neq f(a')$. In other words, a function is injective when distinct inputs always lead to distinct outputs.

A function $f : A \to B$ is *bijective* if it is both injective and surjective, or in other words, if it is both one-to-one and onto. In this case, we say that f puts the sets A and B into a *one-to-one correspondence*. If $f : A \to B$ and $g : B \to C$, then we may define the *composition* function $g \circ f$ as a function from A to C by specifying how it acts on arbitrary input: $(g \circ f)(a) = g(f(a))$. This is indeed a function from A to C, since every $a \in A$ determines the value of $f(a)$, which determines the value of $g(f(a))$, and so for every $a \in A$ there is a unique value for the composition $(g \circ f)(a)$.

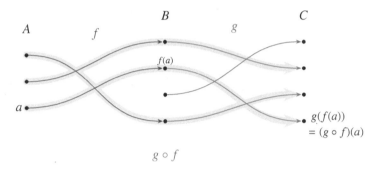

Theorem 94. *The composition of injective functions is injective.*

Proof. Suppose that $f : A \to B$ and $g : B \to C$ are both injective. Consider the composition function $h = g \circ f$, which is a function from A to C. Suppose that $a \neq a'$ are two distinct points in A. Since f is injective, it follows that $f(a) \neq f(a')$ are distinct points in B. Since g is injective, it follows that $g(f(a)) \neq g(f(a'))$, and consequently that $h(a) \neq h(a')$. So h is injective, as desired.   □

If $f : A \to B$ is an injective function, then we may define the *inverse* function f^{-1}, whose domain will be the range of f. If b is in the range of f, then $b = f(a)$ for some $a \in A$, and we define $f^{-1}(b) = a$. This is a function, since the injectivity of f means that a is unique, given b, since no other $a' \neq a$ can have $f(a') = f(a)$. Thus, the object b in the range of f determines the object a in A from which it arose, and so $f^{-1}(b)$ has only one possible value a.

Mathematical Habits

Identify and name the key concepts. Identify the important concepts, and invent precise terminology so that you may recognize and refer to that phenomenon when it arises. It is difficult to reason about an idea when you have no meaningful way to refer to it. So give yourself a way to refer to it by naming it with suitable terminology.

Use words precisely and accurately. Define your terms clearly, and respect those definitions, taking them seriously. Know the meaning of the words you are using. Recognize the definitions of others, which might be different from what you expect. Think like a lawyer about the meaning of the assertions you read or make, as though huge sums or your liberty were at stake. Recognize that words can have a precise technical meaning. My advice to lawyers is, think like a mathematician!

Understand equivalence relations deeply. Use equivalence relations to get at the essence of a concept. Take the quotient of a structure by an equivalence relation, to bring that concept fully to the surface. Understand deeply what it means for an operation or construction to be well defined with respect to an equivalence relation.

Exercises

11.1 Consider the collection of numerical expressions for rational numbers, like $\frac{3}{4}$ or $\frac{-6}{12}$. Let us consider these expressions not as numbers but as syntactic expressions $\frac{p}{q}$, pairs of integers, a numerator p and a nonzero denominator q, so that we count $\frac{1}{2}$ as a different expression than $\frac{2}{4}$. Define the relation $\frac{p}{q} \approx \frac{r}{s}$ for such expressions if they represent the same rational number, which happens precisely when $ps = rq$ in the integers. Prove that this is an equivalence relation.

11.2 Criticize this "proof." Claim. Every transitive symmetric relation R on a set A is also reflexive. Proof. Consider any $a \in A$, and suppose that $a\,R\,b$. By symmetry, we have also $b\,R\,a$. So $a\,R\,b$ and $b\,R\,a$. By transitivity, therefore, $a\,R\,a$, and so the relation is reflexive. $\square$

11.3 Show that if $\sim$ and $\equiv$ are two different equivalence relations on a set A, then the corresponding sets of equivalence classes are not the same.

11.4 Show that congruence modulo n is a congruence with respect to addition and multiplication on the integers, that is, that both addition and multiplication are well defined with respect to congruence modulo n, for every positive integer n.

11.5 Is the squaring function x^2 well defined with respect to congruence modulo 5? How about exponentiation 2^x?

11.6 Criticize this calculation: The residue of the factorial 108! modulo 5 is 1, because 108! mod 5 = (108 mod 5)! = 3! mod 5 = 6 mod 5 = 1.

11.7 Show that the correspondence of equivalence relations with partitions and the correspondence of partitions with equivalence relations provided in the proofs of theorems 92 and 93, respectively, are inverses of each other. That is, if $\sim$ is an equivalence relation on a set X and P is the partition of X arising from $\sim$ in the proof of theorem 92, then the equivalence relation $\sim_P$ arising from P in the proof of theorem 93 is the same as the original relation $\sim$.

11.8 Can a relation be both reflexive and irreflexive? [Hint: Consider whether the set A is empty or not.]

11.9 Can a relation be both symmetric and antisymmetric? [Hint: Suppose that A has only one individual.]

11.10 Prove or refute: All eight combinations of reflexivity, symmetry, and transitivity and their failures are realized in some relations.

11.11 How are the concepts of asymmetric, antisymmetric, and nonsymmetric related to one other? Which imply which others? Which of the eight conceivable combinations actually arise? Provide relations exhibiting each possible combination, and prove any provable implications.

11.12 Give an example of an antisymmetric relation whose reflexive-transitive closure is not anti-symmetric.

11.13 Give a relation on five elements, using as few relational instances as possible, whose reflexive-transitive closure is the full relation. How many relational edges did you use? Make a general claim and argument for a domain with n elements.

11.14 Prove or refute: The symmetric transitive closure of a relation is the same as the symmetric closure of the transitive closure of that relation.

11.15 What is the symmetric closure of the $<$ relation on the integers? What is the reflexive-symmetric closure of the $<$ relation on the integers?

11.16 What is the transitive closure of the *one-less-than* relation M on the integers, that is, the relation $x \, M \, y$, which holds just in case $y = x + 1$? What is the symmetric-transitive closure of this relation?

11.17 What is the symmetric transitive closure of the following relation on the integers: x is either 15 more than or 10 less than y.

11.18 How many equivalence relations are there on a 4-element set? It is possible easily to draw a small picture of each one. Organize this chart into the refinement hierarchy.

11.19 Suppose that equivalence relation E refines equivalence relation F. How do the E-equivalence classes relate to the F-equivalence classes? State and prove a theorem about it.

11.20 Prove that the composition of surjective functions is surjective.

11.21 Prove that the inverse of a bijection from A to B is a bijection from B to A.

11.22 Prove or refute the following claims: If $f : A \to B$ is injective, then (i) $f^{-1} \circ f$ is the identity function on A, and (ii) $f \circ f^{-1}$ is the identity function on B. If the statement is false, state and prove a modified true version of it.

12 Graph Theory

12.1 The bridges of Königsberg

The city of Königsberg, quietly nestled on the banks of the Pregel River, was in the eighteenth century the setting of a mathematical conundrum. The town had seven bridges crossing the river, lovely to traverse in the night air, and the townspeople had a habit of strolling through town in the evening.

The center of the townspeople's mathematical discussions was the pub on the center island, with challenges and late-night attempts to "walk the bridges," to make a tour of the town crossing every bridge exactly once. Despite the accompanying boasts, rarely reproducible in the sober morning, this was a difficult task. Most who attempted found that they had missed a bridge or that they crossed a bridge twice.

Question 95. Can you walk the bridges? Can you find a tour of the town that traverses each of the seven bridges exactly once?

Interlude...

The question had ultimately stumped the town's inhabitants but was finally answered in 1736 by mathematician Leonhard Euler, whose solution gave rise to a new field of mathematics: graph theory. He had in effect invented a new field of mathematics in order to solve the problem. Euler approached the problem by representing it abstractly, finding the mathematical essence of the situation. For this problem, what mattered about the bridges and the town's layout was not their exact shape and location but, rather, the fact that the town had essentially four distinct land masses or neighborhoods—the north bank, the south bank, the central island, and the island at the east—and these were connected by the bridges in a certain way. What mattered was the logical connectivity of those neighborhoods by the bridges.

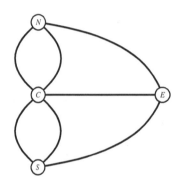

For the purpose of finding a tour, therefore, we may imagine each neighborhood as a single location, a single point or vertex, and each bridge is a line connecting one vertex to another. Thus, the Königsberg bridge problem is represented abstractly by what we now call the *Königsberg graph*, pictured here. More generally, a *graph* is any collection of vertices and edges, where each edge connects one vertex to another or to itself. One can represent a graph visually by drawing the vertices as dots and the edges as lines between them.

12.2   Circuits and paths in a graph

A *path* in a graph is a sequence of vertices in the graph, with each pair of vertices connected by an edge. For graphs that happen to have multiple edges between some of their pairs of vertices, as in the Königsberg graph, we require also that the path should indicate which particular edge was taken at each step, when there was a choice. We may represent each path visually by indicating how the edges are traversed. A graph is said to be *connected* if any pair of vertices admits a path from one to the other. A *circuit* in a graph is a path that starts and ends at the same vertex.

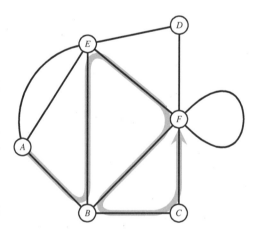

Definition 96. An *Euler circuit* is a circuit that uses every edge in the graph exactly once. An *Euler path* is a path that uses every edge exactly once.

Thus, an Euler circuit is simply an Euler path that starts and ends at the same place. In the figures below, I have indicated an Euler circuit in blue and an Euler path in red.

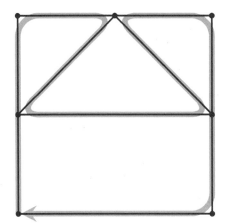 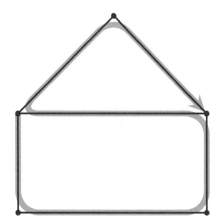

A key concept of Euler's solution of the Königsberg bridges problem is the *degree* of a vertex in a graph, which is the number of times it occurs as the endpoint of an edge. This is the same as the number of edges that touch the vertex, except that we must count any edge from the vertex to itself twice, once for each end separately.

Theorem 97. *A finite connected graph admits an Euler circuit if and only if every vertex has even degree.*

Proof. ($\rightarrow$) Suppose that G is a finite connected graph with an Euler circuit. Fix such a circuit, and consider any vertex v in the graph. The circuit visits v, and each time it does so, it arrives on one edge and leaves on another (or if v is the starting-ending point, then it leaves at the beginning and arrives at the end). So we can pair off the edges attached to v in this way and thereby see that the degree of v is even. So every vertex has even degree.

($\leftarrow$) Conversely, suppose that G is a finite connected graph and every vertex has even degree. Pick any vertex v_0 to be the starting vertex. As a first step, depart from v_0 on any not-yet-used edge, and continue following such not-yet-used edges until you arrive again at v_0. I claim that you will indeed eventually return to v_0. To see this, note that each time you visit another vertex, you use one edge coming in and another going out, and so with every visit, you use an even number of edges at that node. Therefore, there will also be an outgoing edge available, when you arrive at a new node, except for v_0 itself, where we have already used an outgoing edge. So we must eventually land again at v_0. If this circuit is an Euler circuit, then we are done. Otherwise, we may have missed some edges on our path. To handle these, notice that because the graph is connected there must be some vertex w on our path with some unused edges, and so we shall insert a small detour into the middle of our grand circuit. Start on the detour from w, following only edges that are not used on the original grand circuit. Eventually, one must come back to w, by the same reasoning as before. We absorb the detour into the grand circuit, by first following the grand circuit from v_0 to w, and then following the detour from w to w, and then continuing with the grand circuit. By repeating this process finitely many times so as to add additional detours, we thereby construct a circuit from v_0 to v_0 that uses every edge of the graph exactly once. So it is an Euler circuit, as desired.   □

Corollary 98. *There is no Euler circuit in the Königsberg graph. Therefore, there is no walking tour of the bridges in Königsberg traversing each bridge exactly once, while starting and ending in the same location.*

Proof. The Königsberg graph has four vertices, with respective degrees of 3, 3, 3, and 5, all odd. So by theorem 97, there can be no Euler circuit in this graph. Therefore, there can be no walking tour of the town that traverses each bridge exactly once, while starting and ending in the same location.   □

Great! We seem to have solved the Königsberg bridge problem. Or have we? The corollary shows that one cannot successfully walk the bridges in Königsberg by a tour that starts and ends in the same place. This explains why the boasts of success at the pub were never reproducible. But perhaps the townspeople would count it as a successful walking of the bridges if a tour happened to start and end in different parts of town? That is, perhaps the problem concerns Euler paths rather than Euler circuits. Can we prove an analogue of theorem 97 for Euler paths, rather than Euler circuits? Yes, we can.

Theorem 99. *A finite connected graph admits an Euler path if and only if there are at most two vertices with odd degree.*

Proof. ($\rightarrow$) Suppose that G is a finite connected graph with an Euler path. Fix such a path, and observe as in theorem 97 that every vertex v appearing on it, except possibly the initial and terminal vertices, must have even degree, because the path enters each such vertex on one edge and departs on another, using up two edges with each visit. So there are at most two nodes with odd degree, as desired.

($\leftarrow$) One can mount a direct argument for the converse implication here, with an argument similar in style to that of theorem 97. But let us instead deduce this implication more simply as a corollary to theorem 97. We begin with a finite connected graph G in which there are at most two vertices with odd degree. If there are no odd-degree vertices, then we already know by theorem 97 that there is an Euler circuit, and this is also an Euler path. So assume that there is an odd-degree vertex. It follows that there must be exactly two odd-degree vertices v and w, since by exercise 12.2 there cannot be just one. Let G^+ be the graph obtained by adding an edge between the two odd-degree vertices of G. So this is a finite connected graph in which every vertex now has even degree. So by theorem 97 there is an Euler circuit in G^+. By exercise 12.1, we may assume that the circuit ends exactly with the new edge we added between v and w. Thus, if we consider only the part of the circuit before that, we have exactly an Euler path for G, as desired. □

Corollary 100. *There is no Euler path in the Königsberg graph. Consequently, the Königsberg bridge problem admits of no solution.*

Proof. The Königsberg graph has four vertices with odd degree. Therefore, by theorem 99 it admits no Euler path. So there is no tour of the town of Königsberg traversing each bridge exactly once. □

The reader will learn in the exercises that some of the inhabitants of Königsberg seek to address this situation by building more bridges.

12.3 The five-room puzzle

Let us consider next the *five-room* puzzle, which can also be solved using graph theory. Imagine a five-room house, with many doors between adjacent rooms and to the exterior, as indicated in this floor plan from the architect.

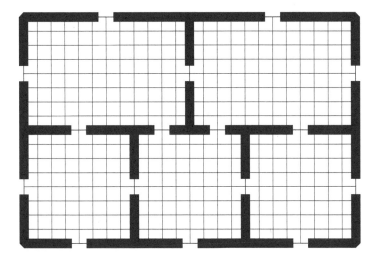

Question 101. Is there a tour of the house that traverses each doorway exactly once?

We assume that one may not pass through the walls except by traversing a doorway and that, once outside, we may freely walk around and enter any other exterior door. Can you find a solution?

Interlude...

Theorem 102. *There is no tour of the five-room house that traverses each doorway exactly once.*

Proof. Consider the graph F that abstractly represents the connectivity of the five-room house. So F has six vertices, one for each room of the house and one vertex for the exterior region, which we can refer to as the "exterior room." The graph F has an edge for each doorway, connecting the vertices associated with each of the rooms that the doorway connects. A tour of the five-room house traversing each doorway exactly once amounts to a Euler path in the associated graph F. By inspection, the upper two rooms each have degree 5 in this graph; the lower three rooms have degree 4, 5, and 4, respectively; and the exterior room has degree 9. Since we have four vertices with odd degree, the graph F admits of no Euler path. And so there is no tour of the five-room house that traverses each doorway exactly once.   □

We can summarize the argument like this: every time you enter a room by one doorway, you must exit it by another, and so except for the starting and ending rooms, they must each have an even number of doors altogether. But they do not, since there are more than two rooms with an odd number of doors. So it is impossible.

A *simple* graph is a graph having no edges from a vertex to itself and also no multiple edges between two of its vertices. Thus, a simple graph is essentially a set with an irreflexive symmetric relation, the edge relation on the vertices.

Theorem 103. *In every finite simple graph with at least two vertices, there are two vertices with the same degree.*

Proof. Suppose that G is a simple graph with n vertices, where $n \geq 2$. Since no vertex has an edge with itself, the degree of any vertex in G is an integer from 0 to $n - 1$, and there are precisely n numbers in that range. But notice that if there is a vertex with degree $n - 1$, then it is connected to all the other vertices, and so in this case there cannot be a vertex of degree 0. (We used our assumption that $n \geq 2$ in order to know that 0 and $n - 1$ are different numbers.) In other words, the degrees 0 and $n - 1$ cannot both arise as the degree of a vertex in G. So there are really only $n - 1$ many possibilities for the degree, and so by the pigeon-hole principle, there must be at least two vertices with the same degree.   □

12.4 The Euler characteristic

Let us consider next the *Euler characteristic*. A finite graph is *planar* if it can be represented by vertices and edges in the Cartesian plane without any crossing edges. Such a graph divides the plane into a finite number of *faces*, the regions bounded by the edges, and we count the region totally outside the graph as one of the faces. The *Euler characteristic* of any such graph is the quantity $v - e + f$, where v is the number of vertices, e is the number of edges, and f is the number of faces.

Theorem 104. *The Euler characteristic of any finite nonempty connected planar nonempty graph is 2.*

$$v - e + f = 2 .$$

Proof. We prove the theorem by induction on the size of the graph or, more specifically, on the number of edges appearing in the graph. The statement is true in the graph with one vertex and no edges. Let us now imagine adding vertices and edges. If we add a single new vertex and a new edge connecting it to a vertex we had before (in order that the graph is connected), then we have increased v and e by 1, but we have not changed f, so we still have $v - e + f = 2$. If we add an edge between two vertices we already have, then this new edge cuts a previous face in two, and so we have increased e and f by 1, but we have not changed v. And so the resulting graph still has $v - e + f = 2$. Since every finite connected planar graph can be constructed by finitely many steps of these kinds, we see that $v - e + f = 2$ in all of them.   □

The reader might enjoy exploring the idea of extending the Euler characteristic to three-dimensional polyhedra, such as cubes, prisms, and other shapes. For example, a cube has $v = 8$ vertices, $e = 12$ edges, and $f = 6$ faces, resulting in the Euler characteristic $v - e + f = 8 - 12 + 6 = 2$. A triangular prism has six vertices, nine edges, and five faces, leading to $v - e + f = 6 - 9 + 5 = 2$, once again. Similarly, a hexagonal prism has twelve

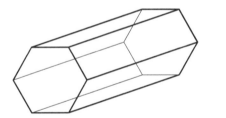 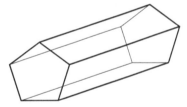

vertices, eighteen edges, and eight faces, leading to $v - e + f = 12 - 18 + 8 = 2$. A pentagonal prism has ten vertices, fifteen edges, and seven faces, leading to $v - e + f = 10 - 15 + 7 = 2$, once again. It works in all these cases and many more—amazing!

Mathematical Habits

Have favorite examples. For any mathematical idea, have a favorite example—a favorite prime number, a favorite polynomial, a favorite discontinuous function, a favorite graph. When learning a new idea, see how it plays out with your favorite examples.

> **Grasp the big picture.** When reading a proof, aim to understand the underlying idea. There are multiple levels involved with reading and understanding proofs. At a basic level, one can read a proof by understanding every word in it and verifying that the steps are all logically correct and that they establish the truth of the theorem. A deeper understanding of a proof, however, involves a holistic understanding of the argument, including the ability to summarize the argument, identifying the main ideas and strategy of the proof. If you have truly grasped a proof, then you can easily state and prove the theorem again entirely on your own.

Exercises

12.1 Prove that if a finite graph admits an Euler circuit, then for any particular edge of the graph, there is such an Euler circuit ending with exactly that edge.

12.2 Prove that in any finite graph, the total degree, meaning the sum of the degrees of all the vertices, is always even. Conclude as a corollary that no finite graph can have an odd number of odd-degree vertices.

12.3 Prove that in any finite graph with exactly two vertices of odd degree, every Euler path must start at one of those odd-degree vertices and end at the other.

12.4 Give a direct proof of the converse implication of theorem 99, analogous to that of the proof of theorem 97. [Hint: Take care to start with the right vertex.]

12.5 Let K_n be the complete graph on n vertices. This is the graph with n vertices and an edge between any two distinct vertices. Show for nonzero n that K_n has an Euler circuit if and only if n is odd. For which n does K_n have an Euler path?

12.6 Suppose that in the city of Königsberg, the North Prince lives in a castle on the north bank and the South Prince lives in a castle on the south bank. The university is on the east island, but the center of the townspeople's mathematical discussions is, as we mentioned, at the pub on the center island, with challenges and late-night attempts to "walk the bridges." Now, the North Prince has a plan to build a new bridge, an eighth bridge, in such a way that from his castle he could traverse all eight bridges and end at the pub, for celebration. Where should he build a bridge that would enable him to do this? Prove that there is essentially only one location for such a bridge, in terms of its connectivity, and furthermore, if such a bridge were built, the South Prince would definitely not have the same ability from his own castle.

12.7 After the North Prince builds his bridge, the infuriated South Prince seeks revenge by building a ninth bridge, which will enable him to traverse all nine bridges exactly once, starting from his own castle and ending at the pub, while simultaneously preventing the analogous ability for the North Prince. Where should he build the ninth bridge?

12.8 In order to settle the dispute—and get more customers—the pub keeper decides to build a tenth bridge, which will enable everyone in town to traverse all the bridges, starting at whichever location they wish. Where must the innkeeper build his bridge?

12.9 Across the street from the five-room house considered in this chapter is another five-room house, pictured below, with a slightly different floor plan. Does this house admit a tour traversing each doorway exactly once? Does it admit such a tour starting and ending in the same place?

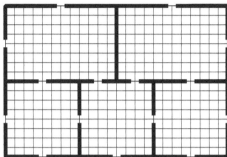

12.10 A graph is *bipartite* if the vertices can be partitioned into two disjoint sets, such that every edge connects a vertex in one set to a vertex in the other set. Prove that if a graph is bipartite, then it has no odd-length cycles. [Hint: An easier case would be to prove that a bipartite graph is triangle-free.]

12.11 The converse is somewhat more challenging: prove that if a graph has no odd-length cycles, then it is bipartite.

12.12 Does the $v - e + f = 2$ conclusion of the Euler characteristic theorem hold for the empty graph? Is this a reason to say that the empty graph is not connected? (The empty graph does not have exactly one connected component, which is sometimes taken as what it means to be connected, although it does have the property, vacuously, that any two vertices are connected by a path.)

12.13 Formulate and prove a version of the Euler characteristic theorem that applies to finite planar graphs that are not necessarily connected. Your formula should involve $v - e + f$ and also refer to the number of connected components of the graph. Does your formula work with the empty graph?

12.14 Prove that every n-gonal prism, the solid with a regular n-gon on two ends and edges connecting the corresponding vertices, has $v - e + f = 2$.

12.15 Construct a triangulation of the torus, and compute the Euler characteristic of it.

12.16 Define that a bidirectional Euler circuit is a circuit in which every edge is traversed exactly twice, once in each direction. Does the Königsberg graph admit a bidirectional Euler circuit? Show that if a graph admits an Euler path, then it admits a bidirectional Euler circuit. Is the converse true? Show that every finite connected graph admits a bidirectional Euler circuit.

Credits

The solution of the Königsberg bridge problem is due to Euler in 1736. The exercises about the eighth, ninth, and tenth bridges appear on the Wikipedia (2020) entry for the Königsberg bridge problem, which also relates the "walking the bridges" storyline. The image of the Königsberg bridges at the opening of the chapter was hand drawn by the author.

13 Infinity

13.1 Hilbert's Grand Hotel

Let me tell you the parable of Hilbert's Grand Hotel. The hotel has infinitely many rooms, each a luxurious full-floor suite, numbered like the natural numbers: 0, 1, 2, and so on, endlessly. Each room has a guest, and the hotel is completely full, with infinitely many guests. But meanwhile, a new guest has just arrived, wanting a room. What is the manager to do?

The manager says coolly, "No problem." He sends an announcement up to all the current guests: everyone must change rooms up to the next higher room.

$$\text{Room } n \quad \mapsto \quad \text{Room } n + 1$$

There was fine print in the agreement that everyone had signed at check-in, you see, stipulating that guests might have to change rooms under the management's direction. So everyone complies, with the result, you will notice, that room 0 now becomes vacant, available for the new guest.

Over the next several days, more guests arrive, and although the hotel is completely full, the manager is able to accommodate them each time by having all the current guests move up again, freeing up an additional room.

On the weekend, a crowd shows up all at once—1000 guests—and the manager accommodates them all in one migration by having all the current guests move up 1000 rooms:

$$\text{Room } n \quad \mapsto \quad \text{Room } n + 1000.$$

This makes available, in one step, all the rooms from 0 to 999, which accommodates the crowd.

Hilbert's bus

The following week, a considerably larger crowd arrives: Hilbert's bus pulls up. The bus has infinitely many seats, numbered like the natural numbers, so there is seat 0, seat 1, seat 2, and so on, and in every seat there is a new guest wanting to check into the hotel. Can the manager accommodate them?

Well, it makes no sense to ask the guests to move up infinitely many rooms, since every individual room has a finite room number, and so he cannot simply move everyone up as before. But can he somehow rearrange the current guests so as to make room for the bus passengers? Please try to figure out a method on your own before reading further!

Interlude...

Did you find a solution? The answer is yes, the guests can be accommodated. The manager simply directs first that the guest currently in room n should move to room $2n$, which frees up all the odd-numbered rooms. And then he directs the passenger in seat s to take room $2s + 1$, which is certainly an odd number, and people in different seats will get different odd-numbered rooms.

$$\text{Room } n \quad \mapsto \quad \text{Room } 2n$$
$$\text{Seat } s \quad \mapsto \quad \text{Room } 2s + 1$$

Thus, everyone is accommodated with a room of their own, the previous guests in even-numbered rooms and the bus passengers in odd-numbered rooms.

Hilbert's train

Now, Hilbert's train arrives.

The train has infinitely many train cars, each with infinitely many seats, and every seat is occupied. The passengers are each identified by two pieces of information: their car number c and their seat number s, and every passenger is eager to check in to the hotel. Can the manager accommodate them all?

Interlude...

Yes, indeed. The manager again directs the resident of room n to move to room $2n$, which again frees up the odd-numbered rooms. And then, the manager directs the passenger in car c, seat s to check into room $3^c 5^s$.

$$\text{Room } n \quad \mapsto \quad \text{Room } 2n$$
$$\text{Car } c, \text{ Seat } s \quad \mapsto \quad \text{Room } 3^c 5^s$$

Since $3^c 5^s$ is certainly an odd number, it is available for a new guest, and different train passengers will take up different rooms because of the uniqueness of the prime factorization. So again everyone is accommodated.

Hilbert's half marathon

Everyone was in town for the big race, and here it comes: Hilbert's half marathon—this crowd likes fractions—with the runners densely packed in amongst themselves. The runners each have a race number, one for every nonnegative rational number. They all want to check into the hotel.

Can the hotel manager accommodate all the runners? Yes. After freeing up the odd-numbered rooms as before, he can place runner p/q in room $3^p 5^q$, where p/q is in lowest terms.

$$\text{Room } n \quad \mapsto \quad \text{Room } 2n$$
$$\text{Runner } \frac{p}{q} \quad \mapsto \quad \text{Room } 3^p 5^q$$

In this way, the entire half marathon fits into Hilbert's hotel. It might be a little surprising at first how similar the manager's solution was for the half marathon to the solution for the train. But ultimately, it should not be too surprising, because every train passenger is determined by two positive integers—the car number and the seat number—and similarly with every marathon runner: the numerator and the denominator.

Cantor's cruise ship

After the race, a shadow falls upon the hotel. It
is Cantor's cruise ship pulling into the harbor,
with a passenger for every real number. There
is passenger $\sqrt{2}$, passenger e, passenger π, and
so on, excited to have finally arrived. Every
ship passenger has a ticket, with a distinct ticket
number printed on it, a real number, and every
real number arises on some ticket. Can we fit
them all into Hilbert's hotel?

CUNARD R.M.S. AQUITANIA   TONNAGE 45,650

13.2   Countability

While you ponder that question on your own,
let us leave the parable and discuss more explicitly the mathematical ideas involved in it.
Although the story may be fanciful, there is serious mathematics here. The main concept at
play is *countability*, for a countable set is essentially one that fits into Hilbert's hotel.

Definition 105. A set is *countable* if it can be placed in one-to-one correspondence with a
set of natural numbers.

In other words, a set A is countable if there is an injective function $r : A \to \mathbb{N}$. In
the Hilbert hotel style, one can conceive of the function r as a room-assignment function,
placing person a in room $r(a)$. The arguments and constructions with Hilbert's hotel above
amount to claims that certain sets were countable.

Another way to think about it is that a nonempty set A is countable if and only if one can
enumerate the elements of A by a list, indexed with the natural numbers

$$a_0, \ a_1, \ a_2, \ a_3, \ \ldots,$$

possibly with repetition (in case the set is finite). This amounts to a surjective function from
$\mathbb{N}$ to A, mapping $n \mapsto a_n$. Given any such mapping, we can produce an injective function
$f : A \to \mathbb{N}$ in the converse direction, by mapping a to the least n for which $a = a_n$.

Theorem 106. *The union of two countable sets is countable. In other words, if A and B
are countable sets, so is $A \cup B$.*

First proof. Assume that A and B are each countable sets. This is just like the Hilbert's bus
situation, if you think of A as the current hotel guests and B as the bus passengers. The set A
can be put into a one-to-one correspondence with a set of natural numbers. Since the entire
set of natural numbers can be put in one-to-one correspondence with the even numbers,

by the association of n with $2n$, we may place A in one-to-one correspondence with a set of even numbers. Similarly, by composing with the map $n \mapsto 2n + 1$, we may place B in one-to-one correspondence with a set of odd natural numbers. In this way, the two correspondences can be performed simultaneously, leading to a one-to-one correspondence of the union $A \cup B$ with a set of natural numbers. $\square$

Second proof. Another way to think about it is that if A is countable and nonempty, then we may enumerate the elements of A as a_0, a_1, a_2, and so on. And similarly, if B is countable and nonempty, we may enumerate the elements as b_0, b_1, b_2, and so on. Thus, we may enumerate the elements of $A \cup B$ simply by interleaving these two enumerations:

$$a_0, b_0, a_1, b_1, a_2, b_2, \ldots$$

Thus, the union set is countable. If either set is empty, then the union reduces to the other one, and so in any case, the union of two countable sets is countable. $\square$

Corollary 107. *The set of integers* $\mathbb{Z} = \{ \ldots, -2, -1, 0, 1, 2, \ldots \}$ *is countable.*

Proof. The set of integers is the union of two countable sets, namely, the natural numbers $\{ 0, 1, 2, \ldots \}$ and the negative integers $\{ -1, -2, -3, \ldots \}$, and is therefore countable by theorem 106. $\square$

One can also see that $\mathbb{Z}$ is countable by enumerating it like this:

$$0, \ 1, \ -1, \ 2, \ -2, \ 3, \ -3, \ \ldots$$

Theorem 108. *There are just as many pairs of natural numbers as natural numbers. In other words,* $\mathbb{N} \times \mathbb{N}$ *is equinumerous with* $\mathbb{N}$. *Indeed, there is a bijective function* $p :$ $\mathbb{N} \times \mathbb{N} \to \mathbb{N}$ *for which* $p(x, y)$ *is a polynomial function in the two variables.*

We shall give several proofs, although the polynomial claim will arrive only with the third proof.

First proof. The first proof is to use our earlier manner of handling Hilbert's train. Namely, we associate any pair of natural numbers (n, m) with the number $3^n 5^m$. This provides a function $d : \mathbb{N} \times \mathbb{N} \to \mathbb{N}$, which is injective because of the uniqueness of prime factorizations. Thus, $\mathbb{N} \times \mathbb{N}$ is in one-to-one correspondence with a set of natural numbers, and so it is countable. $\square$

Second proof. For a second proof, we provide a bijection between $\mathbb{N}$ and $\mathbb{N} \times \mathbb{N}$. The set $\mathbb{N} \times \mathbb{N}$ consists of the lattice points in a grid, as pictured below at the left. Let us imagine taking a walk through this grid on the winding path pictured at the right. The path starts at the origin and then proceeds to zig and zag progressively up and down the diagonals. The main point is that we shall eventually encounter any given lattice point on this path, if we should simply walk long enough. We may associate the initial point on the path with 0, the next with 1, and then 2, and so on; we associate every natural number n with the nth point (x_n, y_n) encountered on this path.

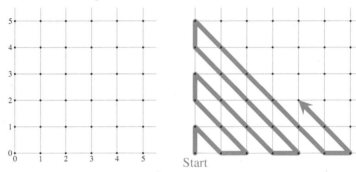

In this way, the function $n \mapsto (x_n, y_n)$ is a one-to-one correspondence between $\mathbb{N}$ and $\mathbb{N} \times \mathbb{N}$.   □

Third proof. A variation of the previous argument will allow us to prove the stronger result, yielding a polynomial bijection $p(x, y)$ between $\mathbb{N} \times \mathbb{N}$ and $\mathbb{N}$. Instead of the winding purple path of the previous proof, we shall instead follow the upward diagonals as pictured here. We start at the origin and then systematically follow each upward diagonal in turn. Let us calculate how many points we have encountered by the time we reach the point (x, y). First, we will have completed a certain number of earlier diagonals, pictured in red. The number of points on these diagonals is 1, 2, 3, and so on. Since each diagonal fulfills the equation $x + y =$ constant, with the constant increasing by one each time

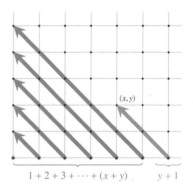

(and one less than the number of points on it), it follows that the final full red diagonal has $x + y$ points. So the total number of points on the red diagonals is the triangular number $1 + 2 + \cdots + (x + y)$, which by theorem 22 (see chapter 4) is equal to $(x + y)(x + y + 1)/2$. Next, the number of points on the final blue line leading up to (x, y) is $y + 1$. Thus, the total number of points up to and including (x, y) is $(x + y)(x + y + 1)/2 + y + 1$. But we want to associate each point with the number of *preceding* points, so that we start with 0 and then continue with 1, 2, and so on, as we walk the path. So the polynomial

$$p(x, y) = \tfrac{1}{2}(x + y)(x + y + 1) + y$$

is a polynomial bijection between $\mathbb{N} \times \mathbb{N}$ and $\mathbb{N}$, as desired.   □

The reader will prove in exercise 13.8 that the integer lattice $\mathbb{Z} \times \mathbb{Z}$ is countable. Meanwhile, the question of whether there is a polynomial bijection between the rational plane $\mathbb{Q} \times \mathbb{Q}$ and the set of rational numbers $\mathbb{Q}$ has been a confounding mathematical question.

Corollary 109. *The set of rational numbers is countable. They can be put into one-to-one correspondence with the natural numbers.*

Proof. We may associate each rational number (p, q), in lowest terms, with the point (p, q) in the integer lattice $\mathbb{Z} \times \mathbb{Z}$. By the result of exercise 13.8, that is a countable set, and so we have associated the fractions with a subset of a countable set. So it is countable. $\square$

For an alternative argument not relying on exercise 13.8, we may associate the positive rational numbers with points in the natural-number lattice $\mathbb{N} \times \mathbb{N}$, which is countable by theorem 108. Similarly, the nonpositive rational numbers can be associated with such points, and so the set of rational numbers is the union of two countable sets, which is therefore countable by theorem 106.

Let us now generalize theorem 108.

Corollary 110. *The countable union of countable sets is countable. In other words, if the sets A_0, A_1, A_2, and so on, are each countable, with A_n for each $n \in \mathbb{N}$, then the union of these sets $\bigcup_{n \in \mathbb{N}} A_n$ is also countable.*

Proof. Since each A_n is countable, there is an injective function $f_n : A_n \to \mathbb{N}$. Let us now associate with any element $a \in \bigcup_n A_n$ the least number n for which $a \in A_n$, and then associate a with the point $(n, f_n(a))$ in the natural number lattice $\mathbb{N} \times \mathbb{N}$. This is a one-to-one correspondence, because from n and $f_n(a)$ we can recover a. So we have an injective function from A to $\mathbb{N} \times \mathbb{N}$, which is bijective with $\mathbb{N}$. So A is countable. $\square$

Another way to understand the proof is that we enumerate the elements of A_0 on the 0-column of $\mathbb{N} \times \mathbb{N}$, the elements of A_1 on the 1-column, and so on. All the elements of the union $\bigcup_n A_n$ are thereby laid out on lattice points of $\mathbb{N} \times \mathbb{N}$. Since $\mathbb{N} \times \mathbb{N}$ is countable, it follows that A itself is countable.

There is an extremely subtle point about the proof of this corollary, which I should like briefly to discuss, although it is a bit beyond the scope of this book; I mention it for those students who will study set theory further. Namely, the argument we gave made use of a certain uniform selection process, when we moved from knowing that each set A_n is countable to having the witnessing injective functions $f_n : A_n \to \mathbb{N}$. In this step, we gave no rule or construction method for selecting a specific such function f_n but, rather, just said that you should choose one, whichever one you like. The general principle asserting that such kind of choices can be made is known as the *axiom of choice*, and deep results in set theory show that there is no way to avoid this kind of choice in the proof of this theorem. Meanwhile, there is a tradition in mathematics to highlight uses of the axiom of choice in mathematical arguments, as I am doing here, and some mathematicians study what set theory can be like without the axiom of choice. In those theories, one cannot say as a general principle that the countable union of countable sets is countable. Meanwhile, in ordinary mathematical arguments it is perfectly acceptable to use the axiom of choice whenever it is convenient, and this axiom is considered to be one of the standard fundamental axioms of set theory.

We may also generalize theorem 108 in the following way.

Theorem 111. *If A is any countable set, then the set A^* consisting of all finite sequences from A is also countable.*

Proof. Since A is bijective with a subset of $\mathbb{N}$, which is bijective with the positive integers $\mathbb{N}^+$, it suffices for us to prove that $(\mathbb{N}^+)^*$, the set of finite sequences of positive natural numbers, is countable. Let us associate to any finite sequence $s = s_0 s_1 s_2 \cdots s_n$ of positive integers the number $2^{s_0} 3^{s_1} 5^{s_2} \cdots p_n^{s_n}$, obtained by placing exponent s_i on the ith prime p_i. This association maps $(\mathbb{N}^+)^*$ into $\mathbb{N}$, and the map is one-to-one because of the uniqueness of the prime factorization of numbers. From any number, we can extract the sequence of exponents from its prime factorization. $\square$

13.3 Uncountability of the real numbers

Let us come now to Cantor's theorem, shocking and profound, showing that the set of real numbers is uncountable. There are different sizes of infinity! Although Cantor's ideas were controversial in his day, sometimes meeting with stubborn rejection, mathematicians eventually recognized his enormous accomplishment. In 1925, Hilbert extolled Cantor's theory of sets and infinite cardinals, announcing, "No one shall cast us from the paradise that Cantor has created for us." Let us see what Hilbert was talking about.

Theorem 112 (Cantor). *The set of real numbers $\mathbb{R}$ is uncountable. It cannot be put into one-to-one correspondence with the natural numbers $\mathbb{N}$, and therefore the size of $\mathbb{R}$ is a higher-order infinity than that of $\mathbb{N}$.*

Proof. Suppose toward contradiction that $\mathbb{R}$ is a countable set, so we can enumerate all the elements of $\mathbb{R}$ as r_1, r_2, r_3, and so on, with r_n for every natural number $n \geq 1$. Every real number appears as some r_n on this list. Using the list, we shall now describe a particular real number z, by specifying its decimal digits

$$z = 0.d_1 d_2 d_3 \cdots$$

Specifically, we fix a decimal representation for each of the real numbers r_n, and we let $d_n = 1$, unless the nth digit of r_n is 1, in which case we set $d_n = 7$. We have now completely specified the real number z. The main point is that we made the nth digit of z different from the nth digit of r_n. (Note that because the digits of z are all either 1 or 7, it has a unique decimal expansion.) It follows that $z \neq r_n$ for any n, and consequently, z does not appear on the list. This contradicts our assumption that every real number appears on the list.   □

Let us illustrate the argument concretely. Suppose that the initial supposed enumeration of the real numbers begins like this:

$$r_1 = \pi$$
$$r_2 = e$$
$$r_3 = \sqrt{2}$$
$$r_4 = \ln(2)$$
$$r_5 = \sqrt{29}$$
$$\vdots$$

We assume that the enumeration continues, eventually listing every single real number exactly once. Let us imagine that we have written out the decimal expansions of these numbers like this:

$$r_1 = 3.\,1\,4\,1\,5\,9\,2\,6\,5\,3\,5\,8\cdots$$
$$r_2 = 2.7\,1\,8\,2\,8\,1\,8\,2\,8\,4\,5\cdots$$
$$r_3 = 1.4\,1\,4\,2\,1\,3\,5\,6\,2\,3\,7\cdots$$
$$r_4 = 0.6\,9\,3\,1\,4\,7\,1\,8\,0\,5\,5\cdots$$
$$r_5 = 5.3\,8\,5\,1\,6\,4\,8\,0\,7\,1\,3\cdots$$
$$\vdots \qquad \ddots$$

In this case, the resulting diagonal real constructed in Cantor's argument above will be

$$z = 0.\,7\,7\,1\,7\,1\cdots$$

whose digits are formed by consulting the digits, highlighted in red, which appear on the diagonal of the array of digits. Specifically, the diagonal construction said to place digit 1 as the nth digit of z, unless the nth digit of r_n is 1, in which case we are to place 7 as the digit. In this way, we ensured that z is different from the first real r_1 in the first digit, that z is different from the second real r_2 in the second digit, and so on. It is different from every r_n, and so z is not on the list, which contradicts our assumption that every real number was on the list.

Cantor's diagonal argument has proved extremely robust, and the diagonal proof method has been used successfully in hundreds or perhaps thousands of mathematical arguments, many of them quite abstract. We shall see a few more diagonal arguments in this chapter. The general feature of a diagonal argument is that one fulfills an infinite list of requirements in an infinitely long construction by satisfying the nth requirement on the nth step.

Let me briefly discuss a subtle aspect of the proof that is sometimes confusing but is ultimately a minor matter. The issue is that not all real numbers have unique decimal expansions. For example, the number one famously has two decimal representations:

$$1 \quad = \quad 1.0000000000\cdots \quad = \quad 0.9999999999\cdots$$

The phenomenon occurs whenever a decimal expansion ends in all 9s or all 0s, for example,

$$1.2468000000\cdots \quad = \quad 1.2467999999\cdots$$

Because of this, it is not always meaningful to speak of *the* decimal expansion of a real number, since some numbers have two. Nevertheless, our diagonal real z used only digits 1 and 7 and therefore has a unique decimal expansion. We are therefore entitled to refer to *the* decimal expansion of z, and furthermore, we need not have done anything special to ensure that $z \neq r_n$, when r_n does not have a unique representation, since this fact itself already ensures that $z \neq r_n$.

Alternative proof of Cantor's theorem

Let me now give an alternative argument of theorem 112, which is closer to how Cantor had first presented his proof, before he had found the argument we gave above.

Alternative proof of theorem 112. Suppose that the set of real numbers is countable. In this case, we can enumerate the real numbers as r_1, r_2, r_3, and so on, a list containing every real number. Let us now describe a certain sequence of numbers $a_n < b_n$, which will be nested, like this:

$$a_1 < a_2 < a_3 < \cdots < b_3 < b_2 < b_1.$$

To start, we consider the real number r_1, and choose $a_1 < b_1$ so that r_1 is excluded from the interval $[a_1, b_1]$. Similarly, at each stage, given the previously specified $a_{n-1} < b_{n-1}$, we shrink this interval to $a_{n-1} < a_n < b_n < b_{n-1}$ in such a way that the next real r_n is excluded from $[a_n, b_n]$.

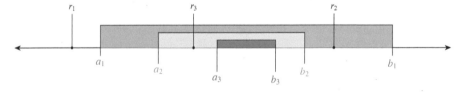

So we have built a nested descending sequence of closed intervals

$$[a_1, b_1] \supseteq [a_2, b_2] \supseteq [a_3, b_3] \supseteq \cdots$$

It follows that the supremum $z = \sup_n a_n$ of the numbers a_n is contained in all the intervals, since it is bigger than any particular a_n and less than any particular b_n. (Corollary 137 in chapter 15 shows more generally that every nested sequence of closed intervals has a common real number inside all of them.) Since every r_n on the list was eventually excluded from the intervals, it follows that $z \neq r_n$ for any n, a contradiction. □

Although the two proofs may look very different at first, one using diagonalization against the digits of the number and the other using a nested sequence of closed intervals, my view is that these arguments are fundamentally the same. They are merely different manners of describing essentially the same underlying construction.

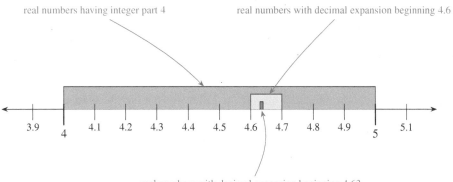

The reason is that to specify the first several digits of a real number in the decimal expansion is exactly to restrict to a certain interval of real numbers, the numbers whose expansion begins with those digits. The resulting intervals are closed intervals, in light of the nonuniqueness of the decimal representations, as with $5 = 4.99999\cdots$, at the right end of the blue interval above. Because of this, the construction in the two formulations of Cantor's argument are essentially identical; the digit diagonalization argument is simply a more concrete and attractive way to describe the particular intervals that are being chosen.

Cranks

Cantor's theorem and proof are sometimes misunderstood in certain characteristic ways, and I sometimes get email, for example, from people who claim that Cantor's proof is all wrong or even that all of mathematics is wrong and so forth. Such messages often come from unfortunate people, who take themselves to be unrecognized geniuses, having refuted what they view as the established dogma of mathematics. If they were correct, then I should be amongst the first to applaud them, since I would give no favor to dogmatism and, furthermore, I would sincerely enjoy any show in which high-and-mighty pompous fools are toppled. But it is simply not the situation here. Upon inspection, one invariably sees, sadly, that the objections they raise are without any force at all. Usually, they have simply misunderstood Cantor's argument.

Let me try to give you a taste of the kind of objections that I have seen raised. In some messages, perhaps with some needlessly complex notation, the person proposes that instead of the Cantor's diagonalization, we should instead define a different diagonal real number y, whose nth digit *agrees* with the nth digit of r_n, rather than disagreeing with it. With this way of arguing, they say, Cantor's argument does not work, and so the real number y could be on the list after all; they conclude that the real numbers could be countable after all. But this is clearly a baseless objection, and the conclusion does not follow. If a modified version of a correct argument does not succeed, it does not mean that the original argument is wrong; it just means that the modified argument does not constitute a proof.

Another objection one sees is that, after having constructed Cantor's diagonal real number z, if it does not appear on the original list r_1, r_2, r_3, and so on, which was supposed to include all real numbers, then one should simply *add it to the list*! Put the diagonal real number z in front, they say, and everything is fine. And furthermore, if one should diagonalize against this new list, then that real also could simply be added to the list. This objection, of course, is without merit, since the claim was not that the diagonal real number z could not appear on any list at all; rather, the claim was that the original list did not already contain all real numbers, since it did not contain the diagonal real z. This contradiction shows that it is impossible to make a list containing all real numbers, and that is precisely what it means to say that the set of real numbers $\mathbb{R}$ is uncountable.

Another common objection is that people sometimes try to prove directly that the set of real numbers is countable. For example, sometimes they start by proving that the set of real numbers with finite terminating decimal expansions is countable. This part is correct, since these real numbers are precisely the rational numbers expressible with a denominator that is a power of 10. Next, they point out that every real number is a limit of a sequence of such rational numbers, which approximate it ever more closely. And that part also is correct. Finally, they attempt to conclude, usually without any proof or with erroneous proof, that this implies that the set of real numbers is countable. But this step of their argument is simply not correct. The set of real numbers does indeed have a countable dense set, but this does not mean that the real numbers themselves are countable.

Shall we leave these silly objections behind and aim for more mathematics?

13.4 Transcendental numbers

Cantor used his theorem to prove the following corollary, giving a new proof of Joseph Liouville's 1844 theorem that transcendental numbers exist, while also strengthening the conclusion. Recall from chapter 1 that a real number is *algebraic* if it is the root of a polynomial with integer coefficients; a number is *transcendental* if it is not. It had been an open question whether there were any transcendental numbers at all, until Liouville's celebrated theorem that they exist. Cantor's goes further, showing that indeed, most real numbers are transcendental.

Corollary 113. *Transcendental real numbers exist. Indeed, most real numbers are transcendental, in the sense that there are only countably many algebraic real numbers but uncountably many transcendental real numbers.*

Proof. A real number r is *algebraic* if it is a root of a nontrivial polynomial in one variable over the integers. Every such polynomial has only finitely many roots, which are naturally ordered by the order of the real numbers. For example, $\sqrt{2}$ is the second root of the polynomial $x^2 - 2$, the first being $-\sqrt{2}$. In this way, every algebraic number is uniquely determined by a finite list of integers: the coefficients of a polynomial for which it is the root and its place among the roots of that polynomial. Since theorem 111 shows that there are only countably many finite sequences of integers, it follows that there are only countably many algebraic numbers.

Meanwhile, theorem 112 shows that the set of all real numbers is uncountable. And so there must be some real numbers that are not algebraic, which is to say that they are transcendental. Furthermore, since the union of two countable sets is countable by theorem 106, there must be uncountably many transcendental real numbers.   □

The proof of corollary 113 gives rise to a subtle philosophical issue. Namely, the question is whether or not Cantor's proof of the existence of transcendental numbers is constructive or not, whether it provides specific real numbers that it proves are transcendental or whether it is instead a pure-existence proof of transcendental numbers. Liouville in 1844 had proved that transcendental numbers exist, after all, by exhibiting a specific real number and proving that it was transcendental. Does Cantor's argument also do this? Does Cantor provide a specific transcendental number? Or is his proof merely a pure-existence proof showing that some real numbers must be transcendental but without exhibiting any particular such real number?

Sometimes mathematicians claim, wrongly in my view, that Cantor's argument is not constructive. The reason, they say, is that Cantor's proof of the uncountability of the set of real numbers proceeds by contradiction—it is uncountable because any proposed enumeration is inadequate. One may then deduce that there must be transcendental numbers, because otherwise there would be only countably many real numbers, which is a contradiction. Indeed, this way of arguing appears to be nonconstructive, since it proves that some real numbers must be transcendental, but it does not seem to exhibit any particular such transcendental number.

But meanwhile, a closer inspection of Cantor's argument shows nevertheless that it *is* constructive, that it does actually provide specific transcendental numbers. First, Cantor provides an explicit enumeration of the algebraic numbers, by means of the finite sequences of integers corresponding to polynomials and the particular roots of each; and second, his diagonalization method provides a way to construct a specific number—the diagonal real—which is not on a given enumeration. By simply combining these two steps, therefore, he has constructed a specific real number that is not algebraic. So Cantor's proof is constructive. Furthermore, by simply varying the manner of choosing the diagonal real (by using different digits than we had chosen before, or by slightly changing the enumeration of the algebraic numbers) we can construct an enormous variety of specific distinct transcendental numbers.

13.5 Equinumerosity

Let us now develop a little more of Cantor's theory of the transfinite. One of the fundamental concepts is that of the equinumerosity relation, aimed at expressing the idea that two sets have the same size. What does it mean precisely to say that two sets have the same size?

Definition 114. We say that sets A and B are *equinumerous*, written $A \simeq B$, if they can be placed into one-to-one correspondence with each other. In other words, $A \simeq B$ if there is a bijection $f : A \to B$, a one-to-one onto function from A to B.

There is a natural partial order underlying equinumerosity, namely, $A \lesssim B$ if there is a one-to-one function $f : A \to B$, not necessarily onto. In exercises 13.9 and 13.10, the reader will verify that the equinumerosity relation $\simeq$ is an equivalence relation and that $\lesssim$ is a partial preorder.

The equinumerosity concept is evidently a more primitive notion of "same size" than the idea of counting the number of elements of the set. We can see easily without counting, for example, that the set of people in the classroom has the same size as the set of noses in the classroom, simply because every person in the room has one nose and there are not any extra noses. So we have constructed a one-to-one correspondence between the set of people and the set of noses, and we have done so without counting these sets.

The equinumerosity concept provides a notion of same size that is fruitfully applicable even to infinite sets.

Theorem 115. *Every countable set is either finite or equinumerous with the whole set of natural numbers $\mathbb{N}$.*

Proof. Since every countable set is equinumerous with a set of natural numbers, it suffices to consider only sets of natural numbers $A \subseteq \mathbb{N}$. If A is infinite, then we may associate the number n with the nth element of A, and this will be a one-to-one correspondence between $\mathbb{N}$ and A, and so A is equinumerous with $\mathbb{N}$. So every set of natural numbers is either finite or equinumerous with the whole set of natural numbers.   □

We say that a set is *countably infinite* if it is countable and infinite, which by theorem 115, is equivalent to being equinumerous with $\mathbb{N}$. Let us now push a little harder with Cantor's idea, extending higher into the transfinite realm.

Theorem 116. *No set X is equinumerous with its power set P(X), the set of all subsets of X.*

Proof. We shall prove in fact that there can be no surjective function from X to $P(X)$, let alone a bijective function. Suppose that $f : X \to P(X)$ is a function. Let

$$A = \{\, x \in X \mid x \notin f(x) \,\},$$

which is a subset of X. If $A = f(a)$ for some $a \in X$, then observe that $a \in A \leftrightarrow a \notin f(a) \leftrightarrow a \notin A$, which would be a contradiction. So A cannot be in the range of f, and so f is not surjective. $\square$

Meanwhile, it is easy to see that $X \lesssim P(X)$, since we may map any element $a \in X$ to its singleton $a \mapsto \{a\}$, which is a subset of X, and this is an injective function. So X is strictly smaller in size than $P(X)$. By simply iterating this process, we thereby produce larger and larger sizes of infinity:

$$\mathbb{N} < P(\mathbb{N}) < P(P(\mathbb{N})) < P(P(P(\mathbb{N}))) < \cdots$$

So there are infinitely many different sizes of infinity. Indeed, for any set A of infinite sets, we may take the union of those sets to get a set at least as large as any of them. By taking the power set, we therefore produce an infinite set that is larger than any set in A. The general conclusion is therefore as follows:

There are more different infinities than any given infinity.

13.6   The Shröder-Cantor-Bernstein theorem

A truly robust theory of equinumerosity is enabled by the following fundamental result.

Theorem 117 (Shröder-Cantor-Bernstein). *If sets A and B are at least as large as each other, then they are equinumerous. Specifically,*

$$\text{if } A \lesssim B \text{ and } B \lesssim A, \text{ then } A \simeq B.$$

In other words, if there is an injective function $f : A \to B$ and another injective function $g : B \to A$, then there is a bijective function $h : A \to B$.

The theorem expresses a highly intuitive fact: whenever two sets are at least as large as each other, then they have the same size. Clearly we want this to be true for our mathematical theory of size, and when the theorem is expressed this way, it might seem obviously true or even trivial. But it is not! This theorem is a deep result. To my way of thinking, it is the theorem itself that entitles us to think of equinumerosity as a concept of "size" in the realm of infinite sets; it is the theorem itself that shows that the concept of relative size is not incoherent in the infinitary context. And this is a fact that requires proof.

Proof. Suppose that $A \lesssim B$ and $B \lesssim A$, so that there are injective functions $f : A \to B$ and $g : B \to A$. We can picture this as follows:

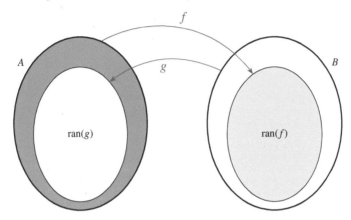

We want somehow to construct a bijective function $h : A \to B$, and we will do so by combining pieces of f and the inverse of g in such a way as to have a one-to-one and onto function h. One of the main issues is that g^{-1} is defined only on the range $\mathrm{ran}(g)$. Thus, if our plan is to succeed, we shall be forced to use f on the part of A that is not in $\mathrm{ran}(g)$, indicated in green above. Therefore, let $A_0 = A - \mathrm{ran}(g)$, and then define recursively

$$A_{n+1} = \{ g(f(a)) \mid a \in A_n \} = g[f[A_n]].$$

These are the sets pictured in green in the figure below. We get A_{n+1} by pushing the set A_n over to B via f to form $f[A_n]$, indicated in orange, and then bringing it back to A via g to form the next set $A_{n+1} = g[f[A_n]]$. So each green stripe gets mapped to a gold stripe,

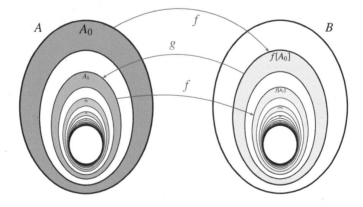

which gets mapped back to the next green stripe in A, in zigzag fashion. Let $A^* = \bigcup_n A_n$

be the union of all the green stripes in A. We finally define the desired function $h : A \to B$ by $h(a) = f(a)$, if $a \in A^*$, and otherwise $h(a) = g^{-1}(a)$. So the proposed bijection h applies f on the green part of A, and it applies g^{-1} on the white part.

I claim that h is a bijection between A and B. To see that it is one-to-one, suppose that $h(a) = h(b)$, but $a \neq b$ are distinct points in A. If $a \in A^*$, then $a \in A_n$ for some n, and in this case $h(a) = f(a)$. Since f is injective, it cannot be that $b \in A^*$, and so $h(b) = g^{-1}(b)$. Since $h(a) = h(b)$, what we have is $f(a) = g^{-1}(b)$, which implies that $b = g(f(a))$, which places b into A_{n+1} and so $b \in A^*$, contrary to our assumption. So $a \in A^*$ is impossible. Similarly, $b \in A^*$ also is impossible, and so neither a nor b is in A^*. In this case, $h(a) = g^{-1}(a)$ and $h(b) = g^{-1}(b)$, which implies that $a = b$ since g^{-1} is injective. So h is one-to-one.

It remains for us to show that h is surjective. Consider any $b \in B$. If $g(b) \notin A^*$, then $h(g(b)) = g^{-1}(g(b)) = b$, which places b into the range of h, as desired. So consider the case $g(b) \in A^*$. Since $g(b) \notin A_0$, as it is in the range of g, it follows that $g(b) \in A_{n+1}$ for some n. But this means that $g(b) = g(f(a))$ for some $a \in A_n$, since every point in A_{n+1} is like this. Since g is injective, it follows that $b = f(a)$, and so $h(a) = b$, which places b into the range of h, as desired. So $h : A \to B$ is a bijection between A and B. □

13.7 The real plane and real line are equinumerous

We may use the theorem to prove the following surprising application.

Theorem 118. *The set $\mathbb{R} \times \mathbb{R}$ is equinumerous with $\mathbb{R}$. Thus, the real plane has the same number of points as the real line.*

Proof. We can clearly embed the line into the plane, and so $\mathbb{R} \leq \mathbb{R} \times \mathbb{R}$. Conversely, given any point in the plane $(x, y) \in \mathbb{R} \times \mathbb{R}$, let z be the real number that interleaves the digits of x and y in their decimal expansions like this:

$$(3.14159 \cdots, 2.71828 \cdots) \mapsto 32.1741185298 \cdots .$$

Some real numbers have two decimal representations, such as with $1.4000 \cdots = 1.3999 \cdots$, although this occurs only for numbers whose representations are either eventually all 9s or all 0s. Therefore, in this interleaving process, let us decide always to use the representations of x and y that do not end in all 9s. The resulting map $(x, y) \mapsto z$ will be injective because from z we can recover x and y by taking every other digit of z (note that z will never end in all 9s). In this way, we have an injection of $\mathbb{R} \times \mathbb{R}$ into $\mathbb{R}$, and so $\mathbb{R} \times \mathbb{R} \leq \mathbb{R}$. This injective function will not be onto $\mathbb{R}$, however, since the nonuniqueness issue means that we shall never realize the number $0.90909090 \cdots$, for example, by interleaving digits this way. Nevertheless, since the sets $\mathbb{R}$ and $\mathbb{R} \times \mathbb{R}$ each inject into the other, it follows by theorem 117 that they are equinumerous as desired. □

In the exercises, you will extend this theorem to higher dimensions. The real line $\mathbb{R}$ is also equinumerous with three-dimensional space $\mathbb{R}^3$ and, indeed, with $\mathbb{R}^n$ in any finite dimension and also with infinite dimensional space $\mathbb{R}^{\mathbb{N}}$.

Mathematical Habits

> **Anthropomorphize.** Understand a difficult idea by perceiving it from the perspective of one of its parts, as though that part were a person, playing a certain role in a structure or organization. Understand how the parts of the problem interact by imagining that they are people who are fighting, when those parts are in tension, or cooperating, when those parts support each other. Find a way to describe a mathematical idea in human terms, in terms of human experience, not in order to make it relevant to humanity but, rather, just to understand better the mathematical idea itself.

> **Use abstraction.** Generalize your arguments via abstraction. Strip away the details of an argument to find the core mathematical idea. Do not fear abstraction—master abstraction by becoming deeply familiar with concrete instances.

Exercises

13.1 Suppose that guests arrived at Hilbert's hotel in the manner described in the chapter: first 6 guests arrive one by one, and then 1000 all at once, and then Hilbert's bus, followed by Hilbert's train, and finally Hilbert's half marathon. If the manager followed the procedure mentioned in the chapter, describe who are the occupants of rooms 0 through 100. How did they arrive and with which party? In which car or seat were they when they arrived, if they arrived by train or bus? Which fraction did they wear in the marathon? If you were the very first guest to arrive, where do you end up in the end? And what is the first room above you that is occupied? How did that guest arrive?

13.2 Down the street from Hilbert's hotel is Hilbert's co-op apartment complex, which is an infinite cubical building, like $\mathbb{N} \times \mathbb{N} \times \mathbb{N}$, where every occupant's residence can be described by a floor number n, a hallway number h, and a corridor r. Because the interior rooms have very little light, the entire cooperative wants to move to Hilbert's hotel. How can the manager accommodate them?

13.3 Consider a dozen or so of the most familiar functions on the real numbers seen in a typical calculus class. Which are functions from $\mathbb{R}$ to $\mathbb{R}$? Which are injective? surjective? bijective?

13.4 Give examples of functions $f : \mathbb{R} \to \mathbb{R}$ that exhibit all possible combinations: bijective, injective not surjective, surjective not injective, neither injective nor surjective.

13.5 Prove that if there is a surjective function $f : \mathbb{N} \to A$, then A is countable.

13.6 Prove or refute: A set A is countable if and only if there is a surjective function $f : \mathbb{N} \to A$. If this is false, state and prove a closely related true theorem.

13.7 Prove that if A and B are sets, with B nonempty, and there is an injective function from B to A, then there is a surjective function from A to B.

13.8 Prove that the integer lattice $\mathbb{Z} \times \mathbb{Z}$ is countable. Can you make several different arguments?

13.9 Prove that the equinumerosity relation $\simeq$ is an equivalence relation.

13.10 Prove that the relation $\lesssim$ as defined in the chapter is a partial preorder: reflexive and transitive.

13.11 Prove that if $A \lesssim B$ and B is countable, then so is A.

13.12 Let $2^{\mathbb{N}}$ denote the set of infinite binary sequences. Give a direct diagonal proof that $2^{\mathbb{N}}$ is uncountable.

13.13 Prove that $2^{\mathbb{N}}$ is equinumerous with the power set $P(\mathbb{N})$.

13.14 Prove that $P(\mathbb{N})$ is equinumerous with $\mathbb{R}$.

13.15 Prove that the set of real numbers $\mathbb{R}$ is equinumerous with the set of complex numbers $\mathbb{C}$.

13.16 Prove that space $\mathbb{R}^3$ has the same number of points as the real line $\mathbb{R}$, and similarly, in any finite dimension that $\mathbb{R}^n$ is equinumerous with $\mathbb{R}$.

13.17 Prove that if A and B are equinumerous, then so are $A \times C$ and $B \times C$ for any set C. Use this to give an inductive proof of the previous exercise.

13.18 Prove that the result of exercise 13.16 remains true in countably infinite dimensions: $\mathbb{R}^{\mathbb{N}}$ is equinumerous with $\mathbb{R}$. [Hint: Since $P(\mathbb{N}) \simeq \mathbb{R}$, it suffices to prove $P(\mathbb{N})^{\mathbb{N}} \simeq P(\mathbb{N})$. But one can prove $P(\mathbb{N})^{\mathbb{N}} \simeq P(\mathbb{N} \times \mathbb{N}) \simeq P(\mathbb{N})$.]

Credits

As a mathematical subject, set theory was introduced by Georg Cantor (1874) in an article including the notions of countability, the uncountability of the real numbers, and the transfinite ordinals. A later article, Cantor (1891), has the diagonal argument explicitly and several of the other generalizations mentioned in this chapter. The Cantor's ship image is by Cunard (Postcards from the Early 1920s), public domain, via Wikimedia Commons. Regarding the issue of whether Cantor's proof of the existence of transcendental numbers is constructive, I was present at a mathematics colloquium lecture where a highly distinguished set theorist made the claim that it was not constructive as part of his talk to a general audience; I spoke to him privately after the talk, and ultimately he agreed with me that Cantor's proof is actually constructive. A discussion concerning the problem whether there is a polynomial bijection of $\mathbb{Q} \times \mathbb{Q}$ with $\mathbb{Q}$ is available on MathOverflow at https://mathoverflow.net/q/21003/1946, including a link to a paper of Bjorn Poonen essentially solving it under a certain well-known number-theoretic conjecture.

14 Order Theory

14.1 Partial orders

Order relations are pervasive, arising in every field of mathematics. We are all familiar with the *less-than* relation $n < m$ in arithmetic or the *less-than-or-equal-to* relation $x \leq y$ in the real numbers. We have the *divides* relation $n \mid m$ in the natural numbers and the *inclusion* relation $A \subseteq B$ for subsets of a given set. We compare growth rates of functions by means of the *eventually dominating* relation $f <^* g$ for real-valued functions on the real numbers $f, g : \mathbb{R} \to \mathbb{R}$. We have the subspace relation for vector spaces, the subgroup relation on groups, and hundreds of other natural examples. Order relations often enable us to express fundamental aspects of the comparisons between our various mathematical objects, by allowing us to say that one thing is part of another or smaller than or simpler than another in a certain precise way captured by the order.

Let us therefore develop a little of the theory of orders and understand more fully the kinds of features that an order can exhibit. To begin, we set out the central dichotomy between the reflexive orders and the strict orders.

Definition 119.

1. A *partial order* on a set A is a binary relation $\leq$ on A that is reflexive, so that $a \leq a$ for every $a \in A$; transitive, so that $a \leq b \leq c$ implies $a \leq c$; and antisymmetric, so that $a \leq b \leq a$ implies $a = b$.

2. A *strict* partial order on A, in contrast, is a relation $<$ on A that is transitive, so that $a < b < c$ implies $a < c$; and irreflexive, so that $a < a$ never holds.

The examples mentioned in the introduction above are each either partial orders or strict partial orders. Can you tell which is which amongst them?

Partial orders and strict partial orders come in pairs, each interdefinable with its partner. For every partial order $(A, \leq)$, we define a corresponding strict order: $a < b$ just in case $a \leq b$ and $a \neq b$. And conversely, for every strict partial order $<$, we may define the corresponding partial order: $a \leq b$ if and only if $a < b$ or $a = b$. The reader will prove in exercise 14.4 that these correspondences are inverses of one another, in the sense that one flips between the same two orders by going from the partial order to the strict order and back again. Because of this, there is a one-to-one correspondence between partial orders and strict orders, and whenever we refer to a partial order $\leq$, we are entitled also to refer to the corresponding strict order $<$, and vice versa.

With partial *pre*orders, in contrast, which are reflexive transitive relations that are not necessarily antisymmetric, this interdefinability correspondence breaks down. This is because there are different preorders $\leq_1$ and $\leq_2$ giving rise to the same strict order $<$. In this sense, with preorders, the reflexive relation $\leq$ has more information. In the subject areas using preorders, therefore, one should take the reflexive preorder as more fundamental than the corresponding strict order.

To aid communication, mathematicians usually use a $\leq$-like symbol for partial orders and a $<$-like symbol for the corresponding strict partial order. For example, one might see symbols like

$$\leq \quad \preccurlyeq \quad \trianglelefteq \quad \subseteq \quad \sqsubseteq$$

used for a partial order, with the corresponding strict partial order denoted with the symbols

$$< \quad \prec \quad \triangleleft \quad \subset \quad \sqsubset.$$

To be clear, in order theory we use these symbols essentially as variables, writing $\leq$, for example, to represent whichever particular order we have in mind in that context. This may be a somewhat more abstract use of this kind of symbol than some students have encountered, which is why I remark specifically on it. In order theory, the symbol $\leq$, say, or $\preccurlyeq$, $\trianglelefteq$, and $\sqsubseteq$, do not necessarily have any fixed, predetermined interpretation or meaning; these symbols represent different specific orders in different contexts, whichever is useful for the argument at hand, just as the symbol x might refer to different real numbers or objects in different arguments.

14.2 Minimal versus least elements

Let us study the following distinction for elements in a partial order.

Definition 120.

1. An element a in a partial order $(A, \leq)$ is a *minimal* element if there is nothing strictly below it, that is, if there is no element b with $b < a$.

2. The element a is a *least* element, in contrast, if it is below all the other elements, so $a \leq b$ for all $b \in A$.

Are these concepts the same? Is a minimal element the same thing as a least element?

Interlude...

No, they are not necessarily the same. Just because you have not eaten anyone, it does not follow that everyone else has eaten you; just because you are not standing on anyone else, it does not mean that everyone else is standing on you. But let us give an explicit example where the concepts are different.

Consider the partial order pictured here. This kind of figure is called a *Hasse* diagram of an order, and one interprets the order as the reflexive transitive closure (as described in chapter 11) of the relation following the lines upward. So this diagram indicates that $a \leq c \leq d$ and $b \leq c$, but a and b are incomparable, not related by the order. In a Hasse diagram, one takes a minimalist approach with a clean but sufficient diagram, drawing the fewest number of edges that are sufficient to indicate the order relation via the reflexive transitive closure. In this order, for example, we do not show direct edges for $a \leq d$ and $b \leq d$, since these can be

deduced by transitivity. The element a is a minimal element, since there is nothing below a, but a is not a least element, since it is not below b. But meanwhile, we do have an implication in the other direction.

Theorem 121. *In any partial order, every least element is also minimal.*

Proof. If a is a least element, then $a \leq b$ for all b in A. If a were not minimal, then what this means is that there would have to be some b with $b < a$. So we would have $b \leq a$ and also $a \leq b$, and therefore $a = b$ by antisymmetry. But then $b < a$ is impossible. So a is minimal. $\square$

Theorem 122. *In any partial order, if a is a least element, then indeed it is the unique minimal element.*

Proof. We have already shown that a least element a is minimal. If b also were minimal, then we would have $a \leq b$, since a is least, but $a < b$ is impossible, since b is minimal. And so $a = b$. $\square$

So least elements, when they exist, are the unique minimal elements of their partial order. Does that property characterize leastness? In other words:

Question 123. If we have a partial order $\langle \mathbb{P}, \leq \rangle$ with a unique minimal element, must that element also be a least element?

This question is actually quite interesting, and I encourage you to put the book aside for a moment and try to answer it. Can you make a counterexample or prove there is none? We seek a partial order with a unique minimal element but no least element.

Interlude...

Welcome back. Did you prove it or find a counterexample? Let us see what we can do.

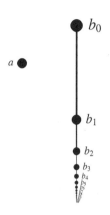

Suppose we have a partial order $\leq$ on a set A with a unique minimal element a. So there is nothing below a. But if a is not least, then there must be some element b_0 that is not above a. So they must be incomparable. But since a is the unique minimal element, then b_0 is not minimal, so there must be some $b_1 < b_0$. And similarly, b_1 is not minimal, and so there is $b_2 < b_1$ and so on. In this way, we are led to construct an infinite chain descending from b_0.

But now, two observations are in order. First, the picture here simply *is* a counterexample! That is, if we have an infinite descending chain and one more point on the side, and no other points at all, then that single point is a unique minimal element, but it is not a least element.

The second observation is that the construction method shows that in any *finite* partial order, every unique minimal element must be a least element, since in a finite order the descending chain part will have to stop at some point, producing another minimal element other than a. These observations essentially prove the following theorem, asserting that the answer to the question above is yes in finite orders but no if infinite orders are allowed.

Theorem 124. *In any finite partial order, an element is a unique minimal element if and only if it is least. Meanwhile, there are infinite partial orders having unique minimal elements that are not least.*

The reader will write a careful proof of this theorem in exercise 14.2.

14.3   Linear orders

Let us now introduce some more vocabulary for discussing other kinds of features in an order. A *linear* order is a partial order $\leq$ with the *linearity* property, which means that for any a and b, we have either $a \leq b$ or $b \leq a$. For example, the usual order $\leq$ on the integers or the rationals or the real numbers is a linear order. A strict order $<$ is said to be linear if $a < b$ or $b < a$ or $a = b$ for any a and b. The reader will prove in the exercises that a partial order is linear if and only if its corresponding strict order is linear.

The subset relation $\subseteq$ is not generally linear, for example, since we can have sets A and B with $A \nsubseteq B$ and $B \nsubseteq A$. For example, if A is the set of prime numbers and B is the set of even numbers, then neither A nor B is a subset of the other.

Theorem 125. *Every infinite linear order has an infinite increasing sequence or an infinite decreasing sequence.*

Let us first prove the following useful lemma.

Lemma 125.1. *If a partial order has no infinite decreasing sequence, then every nonmaximal element a has at least one successor, an element that is minimal amongst the elements above a.*

Proof. If a is not maximal, then there is an element b_0 above a. If this is not a successor of a, then there must be some $b_1 < b_0$, with b_1 still above a. If b_1 is not a successor, then we can find $b_2 < b_1$, but still above a. Since there is no decreasing sequence in the order, this process must terminate at some stage, at which point we will have found a successor of the point a, as desired.   □

Proof of theorem 125. Assume that $\leq$ is a linear order on an infinite set A, and that there is no infinite descending sequence (for otherwise we are done). It follows that there must be a least element a_0 in the order, for otherwise we could build a descending sequence by choosing any b_0, and then any $b_1 < b_0$, and any $b_2 < b_1$ and so on, using the assumption that none of these is minimal. By lemma 125.1, the element a_0 has a successor element a_1. This cannot be a maximal element, since a_0 was least and a_1 was next, and so by linearity all the other infinitely many elements of A are above a_1. So by the lemma again, a_1 has a successor element, which we may call a_2. This also is not maximal, and so has a successor a_3 and so on. At each stage, we have finitely many points at the bottom of the order, but since the order is infinite, the point a_n is not maximal and so has a successor a_{n+1} by the lemma. So we have built an increasing sequence. Thus, every infinite linear order must have either an infinite decreasing sequence or an infinite increasing sequence.   □

In exercise 14.9 the reader will consider whether we really need the linearity assumption in theorem 125. Does every infinite partial order have an infinite increasing or infinite decreasing sequence?

14.4   Isomorphisms of orders

We shall consider next an interesting order arising from the finite sets of natural numbers. Let $P_{\text{fin}}(\mathbb{N})$ be the collection of all finite subsets of $\mathbb{N}$, ordered by the inclusion relation $\subseteq$. This is a partial order. Can you draw a picture of it? Notice that the empty set $\emptyset$ is a least element in this order, and then all the singleton sets $\{n\}$ are minimal above the empty set, with the doubleton sets $\{n, m\}$ minimal above $\{n\}$ and $\{m\}$, and so on. All the finite sets of natural numbers arise in this order.

Let us use this order to illustrate the isomorphism concept in mathematics.

Definition 126. Two order relations $\langle A, \leq_A \rangle$ and $\langle B, \leq_B \rangle$ are *isomorphic* if there is a one-to-one correspondence between A and B, an *isomorphism*, which is a bijective function $\pi : A \rightarrow B$ that copies the $\leq_A$ structure of A to the $\leq_B$ structure of B, in the sense that $x \leq_A y$ if and only if $\pi(x) \leq_B \pi(y)$.

Mathematicians give a lot of respect to their isomorphism concepts. We consider isomorphisms not just between orders but between graphs, between vector spaces, between groups, rings, and fields, and so on. Every mathematical structure gives rise to a corresponding isomorphism concept: two mathematical structures are isomorphic just in case there is an isomorphism between them, a bijective function copying the objects and relations of the first structure exactly to form the second structure. Thus, structures are isomorphic exactly when they are copies of each other, and the isomorphism function details exactly how the copying is carried out, linking each object in one structure with its counterpart in the other. The mathematical attitude toward mathematical structure is that we only ever care about our mathematical structures up to isomorphism—having an isomorphic copy of the structure is just as good, for any mathematical purpose, as having any other copy of it.

Let us illustrate the isomorphism concept by showing that $P_{fin}(\mathbb{N})$ is isomorphic to another familiar order. In chapter 3, we defined that a positive integer is *square-free* if it is not a multiple of any perfect square larger than 1. Equivalently, n is square-free if the prime factorization of n consists of distinct primes, each appearing with exponent one. Let us denote by SF the set of square-free positive integers.

Theorem 127. *The partial order of finite sets of natural numbers $\langle P_{fin}(\mathbb{N}), \subseteq \rangle$ is isomorphic to the partial order of square-free positive integers under divisibility $\langle \mathrm{SF}, | \rangle$.*

Proof. Let us associate every square-free number with the set of primes that divide it. This is a finite set of primes, and because every square-free number is determined by the set of primes that divide it, it follows that different square-free numbers have different prime divisors. Furthermore, division of numbers corresponds to inclusion of these sets of primes, in the sense that $n \mid m$ just in case every prime factor of n is also a prime factor of m. And every finite set of prime numbers arises as the prime factors of the corresponding product, obtained simply by multiplying those primes together. Thus, the association of a square-free number with its set of prime divisors is an isomorphism of $\langle \mathrm{SF}, | \rangle$ with the collection of finite sets of prime numbers $\langle P_{fin}(P), \subseteq \rangle$, where P is the set of prime numbers, ordered by inclusion. Since there are infinitely many primes, we may also see that $\langle P_{fin}(P), \subseteq \rangle$ is isomorphic to $\langle P_{fin}(\mathbb{N}), \subseteq \rangle$, since any bijection between the set of primes P and $\mathbb{N}$ extends to the finite subsets of these sets under inclusion. $\qquad\square$

That was a soft approach, providing a somewhat abstract proof that the two orders are isomorphic. The reader is asked in the exercises to provide an explicit isomorphism between $P_{fin}(\mathbb{N})$ and SF.

14.5   The rational line is universal

Let us turn now to Cantor's remarkable theorem about the rational number order.

Theorem 128 (Cantor). *The rational line $\langle \mathbb{Q}, \leq \rangle$ is universal for all countable linear orders. That is, every countable linear order $\langle L, \leq \rangle$ is isomorphic to a suborder of the rational line.*

Proof. Since L is countable, we may enumerate it as p_0, p_1, p_2, and so on. This enumeration may not necessarily agree with the $\leq_L$ order, and the enumeration may jump around quite a lot in the L order. But that will be fine. We plan to build the isomorphism π from L to a suborder of $\mathbb{Q}$ in stages. Let q_0 be any rational number, and define $\pi : p_0 \mapsto q_0$. This defines the isomorphism on the first point of L.

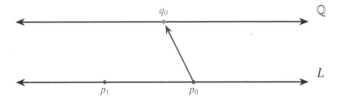

Consider the next point p_1 in L. It is either above p_0 or below it in the L order. We may choose a rational number q_1 that relates to q_0 in just the same way, and define $\pi : p_1 \mapsto q_1$. This defines π on the first two points of L.

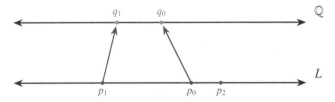

In this way, we gradually associate each point p_n in L with a rational number q_n, such that the map $\pi : p_n \mapsto q_n$ preserves the order. At each stage, the next point p_n is either above all earlier points p_k, below them all, or else in between two of them; and there will always be a corresponding rational number q_n relating to the previous q_k in exactly the same way.

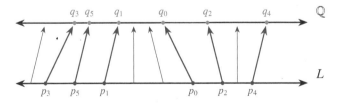

It follows that π is an isomorphism of $\langle L, \leq_L \rangle$ with the corresponding image set $\{ q_0, q_1, q_2, \ldots \}$, which constitutes a suborder of the rational line. So every countable linear order is isomorphic to a suborder of the rational line. □

The key properties of the rational line order $\langle \mathbb{Q}, \leq \rangle$ that we had used in this argument was that it is a linear order; that it is *endless*, which means that it has no maximal or minimal element; and that it is *dense*, which means that whenever $p < r$, there is a point q in between, with $p < q < r$. We need to know those features when we pick the image point q_n, in order to find a counterpart for p_n at stage n.

Theorem 129 (Cantor). *Any two endless countable dense linear orders are isomorphic.*

Proof. This argument is known as the *back-and-forth* argument. We have already seen part of the idea in the proof of theorem 128. Suppose that $\langle L_1, \leq_1 \rangle$ and $\langle L_2, \leq_2 \rangle$ are each countable endless dense linear orders. As before, since the orders are countable, we may enumerate the elements of L_1 as p_0, p_1, p_2, and so on; and we may similarly enumerate the elements of L_2 as r_0, r_1, r_2, and so on. Of course, these enumerations do not necessarily respect the order, and there is no reason to expect that $p_n \mapsto r_n$ is an isomorphism.

We build the desired isomorphism $\pi : L_1 \to L_2$ in stages. At each stage, we will have specified the operation of π on finitely many elements of L_1, in such a way that it is order preserving. Using the method of the previous proof, we may extend the operation of π so as to accommodate any particular p_n. The new point p_n, if not already in the domain of π, is either above all the currently defined members, below them all, or between two of them. Using the fact that L_2 is endless and dense, we may find a corresponding point $r \in L_2$ that relates to the images of the points in just the same way, and define $\pi(p_n) = r$. This is the "forth" step of the back-and-forth argument. For the "back" step, we consider the next element r_n of the target order L_2. If it is not yet placed into the range of π, then it will be either above all the current members of the range of π, or below them all, or between two of them. In each case, we may find a corresponding not-yet-used element p of L_1 and define $\pi(p) = r_n$. In this way, we ensure not only that π is order preserving but also that it will be one-to-one and onto. So the map π that we construct is an isomorphism between L_1 and L_2, as desired. □

14.6   The eventual domination order

Consider next the order of *eventual domination* $f \leq^* g$, which holds of functions $f, g : \mathbb{N} \to \mathbb{N}$ when $f(n) \leq g(n)$ for all *sufficiently large* n, which is a precise compact way of saying that there is a natural number N such that $f(n) \leq g(n)$ for all $n \geq N$. In other words, $f \leq^* g$ holds if eventually $g(n)$ is at least as large as $f(n)$. We may similarly define the strict version $f <^* g$, which means that $f(n) < g(n)$ for all sufficiently large n.

Theorem 130 (Hausdorff). *Every countable set of functions on the natural numbers is bounded above in the eventual domination order. That is, for any countably many functions $f_0, f_1, f_2, \ldots$ on $\mathbb{N}$, there is a function $f : \mathbb{N} \to \mathbb{N}$ that eventually dominates each of them: $f_n \leq^* f$ for every n.*

Proof. Let $f(n) = \sup_{m \leq n} f_m(n)$. Thus, on input n, the value of $f(n)$ is at least as big as $f_m(n)$, for the finitely many f_m with $m \leq n$. It follows that for any particular fixed m, we have $f_m(n) \leq f(n)$ for all $n \geq m$. So $f_m \leq^* f$ for every individual m, as claimed. $\square$

I would like to emphasize how strange and mind expanding this property is. Theorem 130 shows that every countable family in the eventual domination order is bounded. This is a property that is not shared by many of the other natural orders with which you might be familiar. For example, the natural orders on $\mathbb{N}$, $\mathbb{Q}$, and $\mathbb{R}$ do not have this property, since we can easily find unbounded countable sets in these orders. What theorem 130 shows, in contrast, is that it is not possible to construct a ladder of functions

$$f_0 \leq^* f_1 \leq^* f_2 \leq^* f_3 \leq^* \cdots$$

that climbs cofinally or even unboundedly in the eventual domination order. Every such attempt, no matter how high in the order or how fast the functions grow, will still admit an upper bound, a function $f : \mathbb{N} \to \mathbb{N}$ eventually dominating every function f_n in the ladder. It is truly incredible.

Mathematical Habits

Respect isomorphisms. Identify the relevant mathematical structure and incorporate it into your isomorphism concept. Having an isomorphic copy of your structure is just as good, for any mathematical question about that structure, as the original structure.

Exercises

14.1 Prove that if a partial order has two or more minimal elements, then it has no least element.

14.2 Write up a careful proof of theorem 124.

14.3 Provide a partial order with a unique maximal element but no largest element.

14.4 Prove as mentioned in the chapter that the order definitions used in the interdefinability of a partial order with its strict order are inverses of each other. In particular, make a clear statement about what this means. Conclude that the partial orders and strict partial orders are in a one-to-one correspondence.

14.5 Prove the remarks made in the chapter about partial preorders, namely, that there are two different partial preorders $\leq_1$, $\leq_2$ with the same corresponding strict order $<_1 = <_2$. Conclude that with preorders, the $\leq$ relation has more information.

14.6 Prove that a partial order $\leq$ is linear if and only if its corresponding strict order $<$ is linear.

14.7 Prove that the isomorphism relation on partial orders is an equivalence relation.

14.8 Suppose that $\leq$ is a partial order. Define that elements p and q are *comparable* if either $p \leq q$ or $q \leq p$; otherwise, they are *incomparable*. Prove or refute: Comparability is an equivalence relation. Does your answer change if you know that the order is linear?

14.9 Can we omit the linearity assumption in theorem 125?

14.10 Provide an explicit isomorphism $\pi : P_{\text{fin}}(\mathbb{N}) \to$ SF proving theorem 127. [Hint: Use the notation p_n to denote the nth prime number.]

14.11 Prove that the collection of all positive integers $\langle \mathbb{Z}^+, | \rangle$ under divisibility is not isomorphic to $\langle P_{\text{fin}}(\mathbb{N}), \subseteq \rangle$. [Hint: The number 4 has exactly three divisors.]

14.12 Prove that the order of the real line $\langle \mathbb{R}, < \rangle$ is isomorphic to the order on the open unit interval $\langle (0, 1), < \rangle$.

14.13 Consider the collection of all intervals in the real numbers, including open intervals (a, b), closed intervals $[a, b]$, half-open intervals, and rays (a, ∞), $(-\infty, b]$, of all possible open/closed type and bounded/unbounded. Classify all these intervals up to isomorphism. How many different isomorphism types of orders do we find here?

14.14 How does the eventual domination order for functions $f : \mathbb{N} \to \mathbb{N}$ relate to the coordinatewise order, $f \leq g$ just in case $f(n) \leq g(n)$ for every $n \in \mathbb{N}$?

14.15 What is the analogue of theorem 130 for the strict eventual domination order $f <^* g$, for functions $f, g : \mathbb{N} \to \mathbb{N}$?

14.16 State and prove a version of theorem 130 for functions on the real numbers, $f : \mathbb{R} \to \mathbb{R}$. What we want is a notion of domination, "outside a bounded set."

15 Real Analysis

Real analysis is the mathematical subject concerned in its beginnings with the real numbers and with functions on the real numbers. The subject refines and extends many ideas usually first encountered in elementary calculus, including differentiation and integration, but often generalizes them to higher realms, to abstract spaces with infinitely many dimensions or exotic measures. In this chapter, however, we shall remain a little closer to ground, concentrating on the real numbers and continuous real-valued functions on the real numbers.

15.1 Definition of continuity

Perhaps the reader has studied calculus and has experience with the real numbers and with continuous real-valued functions. What does it mean, precisely, for a function to be continuous? In elementary calculus, one is sometimes content at first with an informal concept of continuity. For example, perhaps in high school one might have heard the following account:

A function is continuous if you can draw it without lifting your pencil.

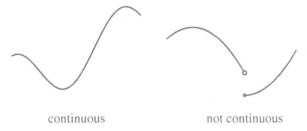

continuous   not continuous

That phrase does suggest that a jump discontinuity, as in the red function at the right, should make a function discontinuous, since indeed you would have to lift your pencil to draw it. But is the meaning of the statement precise enough to serve adequately in a mathematical argument? I do not think so, and in the end I take this phrase as a suggestive metaphor rather than as a definition.

A better definition, but one that in my view remains problematic, is another phrase one commonly hears in an elementary calculus class:

> *A function f is continuous at c if the closer and closer x gets to c, the closer and closer f(x) gets to f(c).*

This phrase is an improvement, since it makes reference to what I find to be the central idea of continuity, namely, the idea that one can obtain increasingly good approximations to the value of a continuous function by applying it to increasingly good approximations to the input. But still, the definition is vague. What exactly does it mean? I claim, furthermore, that a careful reading of this definition will reveal it to be incorrect. Namely, consider the fact that as I walk north from Times Square in New York to Central Park, getting closer and closer to the park, I am also getting closer and closer (ever so slightly) to the North Pole. The problem is that, even though I am getting closer and closer to the North Pole, nevertheless I am not getting *close* to the North Pole, since Central Park is thousands of miles away from the North Pole, and no part of it is close to the North Pole. The suggested definition does not sufficiently distinguish between the idea of getting closer and closer to a quantity and the idea of getting close to it. How close do we want to get? How close suffices? The definition does not say.

Consider the elevation function of a backpacker hiking atop a gently sloped plateau, slowly descending toward the edge, where a dangerous cliff abruptly drops. As she descends the slope toward the cliff's edge, she is getting closer and closer to the edge, and her elevation is getting closer and closer to the elevation of the valley floor (since she is descending, even if only slightly), but the elevation function is not continuous, since there is an abrupt vertical drop at the cliff's edge, a jump discontinuity. Similarly, for the discontinuous red function on the previous page, as x approaches the location c of the jump discontinuity from either side, then for x sufficiently close to c, the value of $f(x)$ does get closer and closer to $f(c)$, even though when approaching from the left, they do not get close to $f(c)$.

For these kinds of reasons, a more correct definition should not refer to "closer and closer" but, rather, should make a precise and explicit mention of exactly how close we want $f(x)$ to be to $f(c)$ and, furthermore, how close it will suffice for x to be to c in order to ensure this. And this is precisely what the usual epsilon-delta definition of continuity achieves:

Definition 131. A real-valued function f defined on an interval or all of the real numbers is *continuous* at a point c if, for every positive $\epsilon > 0$, there is $\delta > 0$ such that every x within δ of c has $f(x)$ within ϵ of $f(c)$. A function is *continuous* if it is continuous at every point in its domain.

One may express the continuity of f at c succinctly in symbols as

$$\forall \epsilon > 0 \; \exists \delta > 0 \; \forall x \quad |x - c| < \delta \implies |f(x) - f(c)| < \epsilon.$$

As we mentioned in chapter 7, the quantifier symbol ∀ here is read as "for all" and the symbol ∃ as "there exists," and so this symbolization asserts that every epsilon has a delta as desired. We can illustrate how these quantities relate in the following figure:

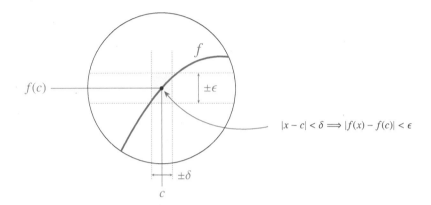

To explain it differently, a function f is continuous at a point c if, for any desired degree ϵ of accuracy, there is a degree δ of closeness to c, such that any x that is that close to c will have $f(x)$ within the desired accuracy of $f(c)$. In short, you can ensure that $f(x)$ is as close as you want to $f(c)$ by insisting that x is sufficiently close to c.

15.2   Sums and products of continuous functions

Let us get a little practice with the definition by proving that the sum of continuous functions is continuous. In the proof, we will use the *triangle inequality*, which asserts that $|u + v| \leq |u| + |v|$ for any real numbers u and v. To prove this, observe that when u and v have the same sign, we have equality, and when they have opposite signs, then we get cancellation on the left but not on the right.

Theorem 132. *The sum of two continuous functions is continuous. In other words, if f and g are real-valued continuous functions on the real numbers, then so is the function $f + g$, defined by addition*

$$(f + g)(x) = f(x) + g(x).$$

Proof. Suppose that f and g are both continuous at a point c, and consider the function $f + g$, whose value at c is $f(c) + g(c)$. To see that this function is continuous at c, fix any $\epsilon > 0$. Thus, also $\epsilon/2 > 0$. Since f is continuous at c, there is $\delta_1 > 0$ such that any x with $|x - c| < \delta_1$ has $|f(x) - f(c)| < \epsilon/2$. And similarly, since g is continuous at c, there is $\delta_2 > 0$ with $|x - c| < \delta_2$ implying that $|g(x) - g(c)| < \epsilon/2$. Let δ be the minimum of δ_1 and δ_2. So if x is within δ of c, then it is both within δ_1 of c and also within δ_2 of c. Consequently, we know that $|f(x) - f(c)| < \epsilon/2$ and also that $|g(x) - g(c)| < \epsilon/2$.

Using the triangle inequality, we may now conclude that whenever x is within δ of c, then

$$|(f(x) + g(x)) - (f(c) + g(c))| = |(f(x) - f(c)) + (g(x) - g(c))|$$
$$\leq |f(x) - f(c)| + |g(x) - g(c)|$$
$$< \frac{\epsilon}{2} + \frac{\epsilon}{2} = \epsilon.$$

We have thereby verified that $f + g$ is continuous at c. $\qquad\qquad\qquad\square$

Let us now prove the analogous theorem for products.

Theorem 133. *The product of two continuous functions is continuous. In other words, if f and g are real-valued continuous functions on the real numbers, then so is the product function $f \cdot g$, defined by multiplication*

$$(f \cdot g)(x) = f(x)g(x).$$

Proof. Suppose as before that f and g are both continuous at a point c, and consider the function $f \cdot g$. Fix any $\epsilon > 0$. Since g is continuous at c, there is some $\delta > 0$ such that any x within δ of c has $g(x)$ within 1 of $g(c)$, and so $|g(x)| < N$ for such x, where $N = |g(c)| + 1$. By taking δ smaller, if necessary, we may also require that $|g(x) - g(c)| < \frac{\epsilon}{2|f(c)|}$ if $f(c) \neq 0$. Thus, for x within δ of c, we will have that

$$|f(c)(g(x) - g(c))| < \epsilon/2,$$

and this will be true whether $f(c) = 0$ or not. Since f is continuous at c, we may ensure, by making δ smaller, if necessary, that $f(x)$ is within $\frac{\epsilon}{2N}$ of $f(c)$.

Putting these facts together, we observe for x within δ of c that

$$|f(x) \cdot g(x) - f(c) \cdot g(c)| = |f(x)g(x) - f(c)g(x) + f(c)g(x) - f(c)g(c)|$$
$$\leq |f(x)g(x) - f(c)g(x)| + |f(c)g(x) - f(c)g(c)|$$
$$= |f(x) - f(c)| \cdot |g(x)| + |f(c)(g(x) - g(c))|$$
$$< \frac{\epsilon}{2N} \cdot N + \frac{\epsilon}{2}$$
$$= \frac{\epsilon}{2} + \frac{\epsilon}{2} = \epsilon.$$

We have therefore verified that $f \cdot g$ is continuous at c. $\qquad\qquad\qquad\square$

Each of the proofs of the previous two theorems involves what is known as an $\epsilon/2$ argument, since the main estimation is made with two pieces, and by controlling the size of each piece to be less than $\epsilon/2$, the whole quantity is kept smaller than ϵ. In analysis, one will find many $\epsilon/2$ arguments, as well as $\epsilon/3$ arguments and even $\epsilon/2^n$ arguments, which make use of the fact that $\sum_{n=1}^{\infty} \frac{\epsilon}{2^n} = \epsilon$ in order to allow a quantity to be controlled in infinitely many separate pieces.

The process of estimating errors is fundamental in real analysis, and the analyst's attitude can be expressed with the slogan:

In algebra, it is *equal, equal, equal.*
But in analysis, it is *less-or-equal, less-or-equal, less-or-equal.*

The point is that, when proving, say, that a function is continuous, you have to bound the size of $|f(x) - f(c)|$, to keep it less than ϵ, and the way that you do this is by showing that it is less than or equal to one thing, which is less than or equal to another, and so on, until you can get ϵ as the final bound. This process is called *estimation.*

The reader will show in exercise 15.1 how theorems 132 and 133 can be used to show that every polynomial function on the real numbers is continuous.

15.3   Continuous at exactly one point

We all know how a collection of good examples can illuminate a mathematical idea, but often having some good nonexamples can also be enlightening, by helping one to see a fuller range of possibilities. So let us consider the following crazy function, which I claim is continuous at exactly one point only and discontinuous at all other points.

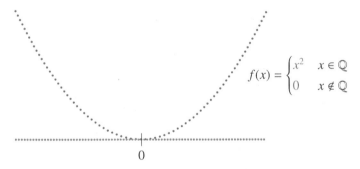

$$f(x) = \begin{cases} x^2 & x \in \mathbb{Q} \\ 0 & x \notin \mathbb{Q} \end{cases}$$

The function is defined essentially as the union of two pieces, by splitting the domain into the rational numbers and the irrational numbers, and doing one thing on rational values and another on irrational values. Specifically, the function $f(x)$ agrees with x^2 when x is rational (shown in blue), but otherwise, it is 0 (shown in red). Despite this bifurcation of cases, the function is actually continuous at 0, since the two separate functions come together at that point. The function is continuous at 0 because for any $\epsilon > 0$, we can ensure that $f(x)$ is within ϵ of $f(0) = 0$ by taking x within $\delta = \sqrt{\epsilon}$ of 0. Meanwhile, the function is discontinuous at all other points $c \neq 0$, since some values of x near c will be very near c^2, but other values of x near c will be 0, and so they cannot all be within ϵ of $f(c)$ if ϵ is smaller than c^2.

In the exercises, the reader will find functions that are discontinuous at exactly two points and that are discontinuous at every point, while the absolute value of the function is continuous.

15.4 The least-upper-bound principle

In order to continue in real analysis, we need to explore a little more explicitly some of the foundational properties of the real numbers, such as the completeness of the real number line. And in order to make this discussion possible, we shall need some terminology. So let us make a few definitions.

A real number r is defined to be an *upper bound* of a set of real numbers $A \subseteq \mathbb{R}$ if every element of A is less than or equal to r. The number r is the *least* upper bound of A, also called the *supremum* of A, if r is an upper bound of A and $r \leq s$ whenever s is an upper bound of A.

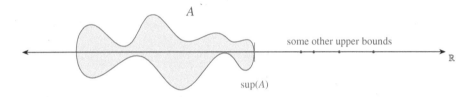

We shall denote the supremum of A, when it exists, by $\sup(A)$. The assertion that every bounded nonempty set of real numbers indeed has a least upper bound is a fundamental principle upon which all the fundamental facts of analysis rest.

Principle 134 (Least-upper-bound principle). *Every nonempty set of real numbers with an upper bound has a least upper bound.*

The least-upper-bound principle, also commonly known as the completeness principle, is to real analysis what the least-number and induction principles are to number theory, used to prove essentially all of the most fundamental properties of these number systems.

15.5 The intermediate-value theorem

In order to illustrate the fundamental nature of the least-upper-bound principle, let us use it to prove some of the familiar basic results of real analysis, beginning with the intermediate-value theorem.

Theorem 135 (Intermediate-value theorem). *If $f : [a,b] \to \mathbb{R}$ is a continuous function on the closed interval $[a,b]$ and d is an intermediate value, meaning that $f(a) < d < f(b)$, then there is a real number c with $a < c < b$ and $f(c) = d$. So every intermediate value is realized.*

Proof. Consider the set A consisting of all real numbers x in the interval $[a, b]$ for which $f(x) < d$. This set is nonempty, because a is in A, and it is bounded, because it is contained in $[a, b]$. So by the least-upper-bound principle, it has a least upper bound, or supremum.

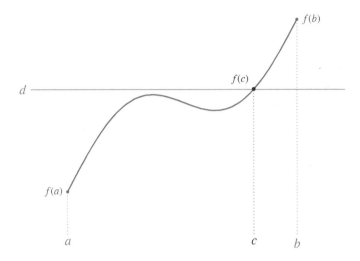

Let $c = \sup(A)$ be the supremum of A, and consider the value of $f(c)$. If $f(c) < d$, then by the continuity of f, the values of $f(x)$ for x near c will all also be less than d (why?). But this would mean that A has some elements above c, contrary to $c = \sup(A)$. Therefore, $f(c) \geq d$. Similarly, if $f(c) > d$, then by the continuity of f, the values of $f(x)$ for x in a small neighborhood of c will all be above d (why?). This contradicts $c = \sup(A)$, since there must be elements of A as close as desired to c, and those elements x must have $f(x) < d$ by the definition of A. Therefore, it must be that $f(c) = d$, and we have found the desired c realizing the intermediate value d. □

Several questions were raised in the proof, which the reader will answer in exercise 15.6, using the precise nature of the epsilon-delta definition of continuity.

15.6 The Heine-Borel theorem

Next, we prove the Heine-Borel theorem. For someone new to real analysis, perhaps the importance of the conclusion of this theorem may not be readily apparent. Why would it be important that every open cover of the closed unit interval admits a finite subcover? Nevertheless, mathematicians have realized that the result has fundamental importance, and the finite-subcover property, it turns out, has an enormous number of consequences in the subject. This compactness property lies at the core of hundreds of arguments and gives rise to the subject known as point-set topology. Let us have a small taste of it in this chapter.

Theorem 136 (Heine-Borel theorem). *If $\mathcal{U}$ is a set of open intervals covering the closed unit interval in the real numbers $[0, 1]$, then there are finitely many open intervals of $\mathcal{U}$ that already cover $[0, 1]$.*

Proof. Fix the open cover $\mathcal{U}$, consisting of open intervals in the real numbers, such that every number $x \in [0, 1]$ is in some $U \in \mathcal{U}$. Consider the set B consisting of all $x \in [0, 1]$ such that the interval $[0, x]$ is covered by finitely many elements of $\mathcal{U}$. Thus, $0 \in B$, since $[0, 0] = \{0\}$ has only one element, and this is in some $U \in \mathcal{U}$, so $\{0\}$ is covered by just one element of $\mathcal{U}$.

Let b be the least upper bound of B in $[0, 1]$. If $b \geq 1$, then we can cover $[0, 1]$ with finitely many elements of $\mathcal{U}$, and we are done. So assume that $b < 1$. Since $\mathcal{U}$ covers $[0, 1]$, there is some interval $U \in \mathcal{U}$ with $b \in U$. Fix this U, which must have the form (r, s). So we have $r < b < s$. Since $r < b$, there must be some $a \in B$ with $r \leq a < b$. Since $a \in B$, there are finitely many $U_0, \ldots, U_n \in B$ with $[0, a] \subseteq U_0 \cup \cdots \cup U_n$. Since $[0, b] \subseteq [0, a] \cup (r, s)$, it follows that $[0, b]$ is covered by $U_0 \cup \cdots \cup U_n \cup U$. And since $b < s$, this union also covers $[0, b']$ for any $b' < s$. So every such b' is in B, including those that are larger than b, contradicting the assumption that b is the least upper bound of B. □

The reader will generalize to the case of arbitrary closed intervals $[a, b]$ in exercise 15.11.

Corollary 137. *Every nested descending sequence of closed intervals has a nonempty intersection. That is, if*

$$[a_0, b_0] \supseteq [a_1, b_1] \supseteq [a_2, b_2] \supseteq \cdots$$

then there is a real number z inside every interval $z \in [a_n, b_n]$.

Proof. Assume that $[a_0, b_0] \supseteq [a_1, b_1] \supseteq [a_2, b_2] \supseteq \cdots$ is a nested descending sequence of closed intervals. Let $z = \sup_n a_n$. So $a_n \leq z$ for every n, and since every b_k is an upper bound of all a_n, it follows also that $z \leq b_k$ for all k. So z is in every interval $[a_n, b_n]$, as desired. □

One can mount an alternative proof of the corollary, based on compactness, and this proof applies generally to closed sets, not just intervals. Namely, if a real number z is not in all the intervals, then there is some n such that $z \notin [a_n, b_n]$, and in this case, we may find an open interval I_z containing z but disjoint from that interval $[a_n, b_n]$ and hence also disjoint from all later intervals, because they are nested. If there is no z in the main intersection set, therefore, the collection of open intervals I_z covers $[a_0, b_0]$, and so by theorem 136 (generalized to arbitrary closed intervals) there are finitely many such intervals $I_{z_1} \cup I_{z_2} \cup \cdots \cup I_{z_k}$ that already cover $[a_0, b_0]$. But each such interval I_{z_i} is disjoint from the intervals $[a_n, b_n]$ for all $n \geq N_i$ for some N_i. So if $n \geq \max(N_1, N_2, \ldots, N_k)$, therefore, the interval $[a_n, b_n]$ is disjoint from each of the I_{z_i}, and so they do not cover $[a_0, b_0]$ after all, a contradiction. So there must be some z common to all the intervals, as desired.

A real number r is a *cluster point* of a set A if every open interval $(r - \epsilon, r + \epsilon)$ about r contains a point from A other than r. In other words, elements of A other than r can be found as close to r as desired.

Corollary 138. *Every infinite bounded set of real numbers has a cluster point.*

Proof. Suppose that A is a set of real numbers contained in the interval $[a, b]$. If some r is not a cluster point of A, then there is an open interval I_r containing r and containing no other elements of A. So if $r \in A$, then I_r has exactly one element of A, and otherwise none.

If the set A has no cluster points, then the union of these intervals covers $[a, b]$. By the Heine-Borel theorem for the interval $[a, b]$, it must be that finitely many of these intervals I_r cover $[a, b]$. But since each interval contains at most one element of A, it follows that A is finite, contrary to our assumption. □

15.7 The Bolzano-Weierstrass theorem

We define that a sequence $\langle a_0, a_1, a_2, \dots \rangle$ of real numbers is *monotone* if either it is monotone *nondecreasing*, meaning that $a_0 \le a_1 \le a_2$ and so on, or it is monotone *nonincreasing*, meaning that $a_0 \ge a_1 \ge a_2$ and so on. Mathematicians are often a little fussy about distinguishing between an *increasing* sequence, which would mean $a_0 < a_1 < a_2$ and so on, strictly increasing at each step, and a monotone *nondecreasing* sequence, which is the weaker property that we mentioned, $a_0 \le a_1 \le a_2$ and so on. There is a similar difference between *decreasing* and monotone *nonincreasing*.

Theorem 139 (Bolzano-Weierstrass theorem). *Every sequence of real numbers has a monotone subsequence. Consequently, every bounded sequence of real numbers has a convergent subsequence.*

Proof. The theorem is related to theorem 125 (see chapter 14). Suppose that $\langle a_0, a_1, a_2, \dots \rangle$ is a sequence of real numbers. Let us define that a term a_n in the sequence is a *viewing point* if $a_n > a_m$ for all $m > n$.

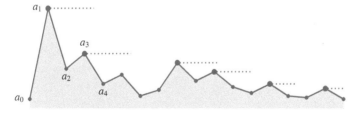

That is, a_n is a viewing point if from it one has an unobstructed view, as though from a mountain outcrop, overlooking all the subsequent terms of the sequence, all the way to the horizon, as indicated here for some of the points in the diagram. If it happens that there are infinitely many viewing points, then there will be an infinite strictly decreasing sequence, since the viewing points themselves form such a sequence: each next viewing point must be strictly lower than the previous one. So in that case, we are done.

Alternatively, consider the case that there are only finitely many viewing points. In this case, pick any a_{n_0} beyond all of them. Since it is not a viewing point, there is some later term with $a_{n_0} \leq a_{n_1}$. And since a_{n_1} is not a viewing point, there is a further term $a_{n_1} \leq a_{n_2}$, and so on. In this way, we find a monotone nondecreasing subsequence $\{a_{n_i}\}$, as desired. So in any case, there is a monotone subsequence.

It follows that every bounded sequence of real numbers has a convergent subsequence, since one may first pass to the monotone subsequence and then observe that every bounded increasing sequence converges to its supremum and that every bounded decreasing sequence converges to its infimum. □

Can one improve the theorem to show that every sequence has a strictly increasing or strictly decreasing subsequence? No, of course not, because it might be eventually constant. But in exercise 15.14, the reader will explain how to strengthen the conclusion of the theorem to strictly monotone subsequences.

15.8 The principle of continuous induction

Earlier I made an analogy between the principle of mathematical induction in number theory and the least-upper-bound principle in real analysis. This analogy is stronger than may be supposed, in light of the principle of continuous induction (see below), an induction-like formulation of the least-upper-bound principle, which can be used to derive many fundamental results in real analysis.

Theorem 140 (Continuous induction principle). *If B is a set of nonnegative real numbers such that*

1. $0 \in B$;

2. whenever $x \in B$, then there is some $\delta > 0$ with $[x, x + \delta) \subseteq B$; and

3. whenever x is a nonnegative real number and $[0, x) \subseteq B$, then $x \in B$,

then B is the set of all nonnegative real numbers.

One can view statement (1) as an instance of (3), if you interpret $[0, 0)$ as the empty set, which would satisfy the hypothesis of (3) vacuously. This is exactly analogous to the fact that the principle of strong induction on the natural numbers does not need to be anchored.

Proof. Suppose toward contradiction that B does not contain all nonnegative real numbers. So there is some nonnegative real number d that is not in B. Let x be the supremum of the numbers a for which $[0, a] \subseteq B$. Since $0 \in B$, we know that $0 \leq x$. Also, $[0, x) \subseteq B$, and so $x \in B$. But then, there is some positive $\delta > 0$ with $[x, x + \delta) \subseteq B$. But in this case, the supremum defining x would have been bigger, a contradiction. □

For a similar principle, Chao Yuen Ren, Chao (1919) puts it like this:

> The theorem is a mathematical formulation of the familiar argument from "the thin end of the wedge," or again, the argument from "the camel's nose."
>
> Hyp. 1. Let it be granted that the drinking of half a glass of beer be allowable.
>
> Hyp. 2. If any quantity, x, of beer is allowable, there is no reason why $x + \delta$ is not allowable, so long as δ does not exceed an imperceptible amount Δ.
>
> Therefore, any quantity is allowable.

Chao's principle, however, is weaker than the principle of continuous induction as we have stated it, because in his formulation, every number has a uniform amount Δ of increase still in the set, whereas with our version of the principle, δ can depend on x. This difference makes Chao's account much closer to induction on the natural numbers, which is how he describes and proves it, whereas our version interacts with (and is equivalent to) the completeness property on the real numbers. Notice also that Chao's version does not require an analogue of condition (3).

The reader will prove some alternative versions of the principle of continuous induction in exercises 15.16 and 15.17, including a version for the entire real number line and a version for closed intervals $[a, b]$.

I mentioned that the principle of continuous induction can be used as a foundational principle from which one can develop essentially all of the basic results in real analysis. Let me spend the rest of this chapter giving some examples of this phenomenon.

Theorem 141. *If $f : [a, b] \to \mathbb{R}$ is a continuous function on a closed interval, then the range of f is bounded.*

Proof. Let B be the set of x for which the range of f on $[a, x]$ is bounded. This contains the point a. If $x \in B$, then we can increase it slightly while staying in B, unless x is already b itself. And if B contains $[0, x)$, then using continuity at x, we can see that f is still bounded on $[0, x]$, and so $x \in B$. So B contains every real number in $[a, b]$. Thus, the range of f on $[a, b]$ is bounded. $\square$

Corollary 142. *Every continuous function $f : [a, b] \to \mathbb{R}$ on a closed interval attains its maximum.*

Proof. By theorem 141, the range of f is bounded. So it has a least upper bound r. If there is no x in the interval $[a, b]$ with $f(x) = r$, then we have $f(x) < r$ for all such x. Thus, $r - f(x)$ is never zero on this interval, and so $1/(r - f(x))$ is a continuous function on $[a, b]$. This function is bounded, and so there is a number N with $1/(r - f(x)) \leq N$ for all x in $[a, b]$. It follows that $r - f(x) \geq 1/N$ and, consequently, that $f(x) \leq r - 1/N$ for all such x. This contradicts our assumption that r was the least upper bound of the range of f on $[a, b]$. Therefore, there must be some x in $[a, b]$ with $f(x) = r$, as claimed. $\square$

Next, let us prove the intermediate-value theorem via continuous induction.

Theorem 143 (Intermediate-value theorem). *If $f : [a, b] \to \mathbb{R}$ is a continuous function on the closed interval and $f(a) < d < f(b)$ for some real number d, then there is a real number c with $a < c < b$ and $f(c) = d$.*

Proof. Let B be the set of $x \in [a, b]$ with $f(x) < d$. Thus, $a \in B$, and if $x \in B$, then by continuity, we know that $[x, x + \delta) \subseteq B$ for some positive $\delta > 0$. Since $b \notin B$, it follows that B cannot satisfy the third hypothesis of the continuous induction principle. So there must be some x with $[0, x) \subseteq B$ but $x \notin B$. In this case, it cannot be that $f(x) > d$, and it also cannot be that $f(x) < d$. So $f(x) = d$, as desired.   □

At the chapter's beginning, we defined what it means for a real-valued function f to be continuous. Our epsilon-delta formulation admits natural variations, which give rise to somewhat stronger conceptions of continuity.

Definition 144. A real-valued function f defined on an interval in the real numbers or on all the real numbers is *uniformly continuous* if, for every $\epsilon > 0$, there is $\delta > 0$ such that whenever points x and y in the domain of f are within δ of each other, then $f(x)$ and $f(y)$ are within ϵ. In symbols, this is expressed by

$$\forall \epsilon > 0 \; \exists \delta > 0 \; \forall x, y \quad |x - y| < \delta \implies |f(x) - f(y)| < \epsilon.$$

Do you see how this is different than mere continuity? The difference is that, in order for f to be continuous, what we need is that, at each point c separately, every ϵ gets a δ. But with uniform continuity, what we need is that every ϵ gets a δ that works with all points in the domain. For continuity, δ can depend on the point c, but with uniform continuity, the same δ must work uniformly for all points.

Theorem 145. *If f is a continuous real-valued function on a closed interval $[a, b]$, then f is uniformly continuous on $[a, b]$.*

Proof. Fix any $\epsilon > 0$. We want to find a $\delta > 0$ so that $|x - y| < \delta \implies |f(x) - f(y)| < \epsilon$ for all x, y in the domain of f. Let B be the set of r such that there is such a δ that works for all x, y in the interval $[a, r]$. In particular, a is in B, since the interval $[a, a]$ has only one point. Next, suppose that $r \in B$, so that there is some δ_0 that works for all points in $[a, r]$. Since f is continuous at r, there is δ_1 such that any x within δ_1 of r has $f(x)$ within $\epsilon/2$ of $f(r)$. Let $\delta = \min(\delta_0, \delta_1)/2$, and consider any x, y in the interval $[a, r + \delta]$, with $|x - y| < \delta$. If both x and y are less than r, then by the choice of δ_0 we know that $|f(x) - f(y)| < \epsilon$, as desired. Otherwise, one of them is larger than r. But since $\delta \leq \delta_1/2$, it means that both x and y must be within δ_1 of r. In this case,

$$|f(x) - f(y)| = |f(x) - f(r) + f(r) - f(y)|$$
$$\leq |f(x) - f(r)| + |f(r) - f(y)|$$
$$< \frac{\epsilon}{2} + \frac{\epsilon}{2} = \epsilon.$$

So δ works for all x, y in the interval $[a, r + \delta]$, and so $[r, r + \delta] \subseteq B$.

Finally, suppose that $[a, r) \subseteq B$. By the continuity of f at r, there is δ_1 such that any x within δ_1 of r has $f(x)$ within $\epsilon/2$ of $f(r)$. And since $[a, r - \delta_1/2] \subseteq B$, we get a δ_0 that works for all x, y in $[a, r - \delta_1/2]$. So $\delta = \min(\delta_0, \delta_1)/2$ will work for all x, y in $[a, r]$, since if x and y are both in $[a, r - \delta_1/2]$, then $|f(x) - f(y)| < \epsilon$ by the choice of δ_0; and otherwise, both x and y are within $\delta_1/2$ of r, in which case the $\epsilon/2$ argument above shows again that $|f(x) - f(y)| < \epsilon$. So we have proved that B is an inductive set, meaning that it fulfills the requirements of the principle of continuous induction, and so by that principle, it follows that $B = [a, b]$. So $b \in B$ and therefore there is a δ that works with ϵ for all x, y in $[a, b]$. So f is uniformly continuous on this interval. $\square$

Mathematical Habits

Write with clarity. Use plain language when possible in preference to technical jargon or symbols. Rely on technical language or detailed notation only when this actually adds clarity. Express explicitly the motivating idea behind your method. Write and rewrite; edit your writing to improve it again and again, especially the summative parts. Explain the key ideas well to cultivate mathematical insight in your reader.

Follow sound mathematical writing conventions. Do not start a sentence with a mathematical symbol. Do not allow mathematical expressions to be separated in the text only by punctuation lest they be interpreted as a single confusing mass of mathematical notation. To improve understanding, insert guiding descriptive words into your sentences that name the types of your mathematical objects. For example, write "The function f is continuous at the point x" instead of just "f is continuous at x" for greater clarity.

Exercises

15.1 Prove that the identity function and every constant function on the real numbers is continuous. Using theorems 132 and 133, prove by induction on the degree that every polynomial function is continuous.

15.2 Prove or refute: If f and g are real-valued functions, then $f + g$ is continuous if and only if f and g are each individually continuous. Prove or refute: $f \cdot g$ is continuous if and only if f and g are individually continuous.

15.3 Find a function $g : \mathbb{R} \to \mathbb{R}$ on the real numbers that is discontinuous at every point except two, the points a and b.

15.4 Find a function $h : \mathbb{R} \to \mathbb{R}$ on the real numbers that is discontinuous at every point but whose absolute value $|h|$ is continuous at every point.

15.5 Suppose that we had defined continuity of a function f using $\le \epsilon$ and $\le \delta$ in place of $< \epsilon$ and $< \delta$, respectively. Would it be an equivalent definition?

15.6 Show that if f is a real-valued function continuous at c and $f(c) < d$, then there is $\delta > 0$ such that every x in the interval $(c - \delta, c + \delta)$ has $f(x) < d$. State and prove a similar theorem for the case $f(c) > d$.

15.7 Prove that every bounded nonempty set A of real numbers has a greatest lower bound.

15.8 Does the empty set have an upper bound in the set of real numbers? Does it have a least upper bound?

15.9 Does the least-upper-bound property hold for the rational numbers?

15.10 Prove that every triangle in the plane has a line that bisects both the area and the perimeter simultaneously. [Hint: Show by the intermediate-value theorem that every line angle yields a unique line that bisects the area of the triangle; show that the ratio of the resulting perimeter partition is continuous in the angle of the line; by rotating the angle through 180°, conclude by the intermediate-value theorem that there is a line bisecting both area and perimeter.] Is it necessarily unique?

15.11 Generalize the Heine-Borel theorem to arbitrary closed intervals $[a, b]$ in the real numbers. That is, prove that if $\mathcal{U}$ is a set of open intervals covering $[a, b]$, then there are finitely many intervals in $\mathcal{U}$ that cover $[a, b]$.

15.12 Prove or refute the analogue of corollary 137 for descending nested sequences of open intervals.

15.13 Give an alternative proof of the Heine-Borel theorem using the principle of continuous induction.

15.14 Prove a slight strengthening of the Bolzano-Weierstrass theorem: every infinite sequence of real numbers contains either a strictly decreasing subsequence, a strictly increasing subsequence, or a constant subsequence.

15.15 In the text, we used the least-upper-bound principle to prove the continuous induction principle. Prove the converse. That is, assume the principle of continuous induction, and derive the least-upper-bound principle as a consequence.

15.16 Prove the following version of the continuous induction principle: Suppose that $B \subseteq \mathbb{R}$ is a nonempty set of real numbers and that (1) whenever $r \in B$, then there is some $a < r < b$ with the open interval $(a, b) \subseteq B$; and (2) whenever $a < b$ and $(a, b) \subseteq B$, then there are r and s with $r < a < b < s$ and $(r, s) \subseteq B$. Then $B = \mathbb{R}$ is the entire set of real numbers.

15.17 Prove the following continuous induction principle for closed intervals: Suppose that $a < b$ and that $B \subseteq [a, b]$ has the properties (1) $a \in B$; (2) whenever $r \in B$ and $r < b$, then there is $\delta > 0$ with $[r, r + \delta) \subseteq B$; and (3) whenever $[0, r) \subseteq B$, then $r \in B$. Then $B = [a, b]$ is the entire closed interval.

15.18 Prove that the function $f(x) = x^2$ is not uniformly continuous on the real numbers $\mathbb{R}$.

Credits

I was put on to the principle of continuous induction by Pete Clark, whose excellent manuscript, Clark (2019), includes an account of this principle and many interesting applications, as well as a history of the idea, which evidently goes back at least to Aleksandr Khinchin in 1922 and the earlier instance with Chao (1919) that I mentioned. Chao's version probably should not be considered a true version of continuous induction, however, because of his uniform-increase requirement and the lack of equivalence with the completeness principle.

Answers to Selected Exercises

In order to provide some helpful guidance to students working the exercises in this book, to show what an acceptable answer might look like, I have written here some sample answers to a few of the exercises, one from each chapter. In each case, I have followed the theorem-proof format, formulating a theorem statement from the instructions in the exercise and then embarking on a proof of that theorem. A fuller collection of answers to the exercises in this book is available as a separate volume from MIT Press.

1.1   Prove that the square of any odd number is odd.

Theorem. *If n is an odd integer, then n^2 also is odd.*

Proof.   Assume that n is an odd integer. By definition, this means that $n = 2k + 1$ for some integer k. In this case, $n^2 = (2k + 1)^2 = (2k + 1)(2k + 1) = 4k^2 + 4k + 1$. We can write this as $2(2k^2 + 2k) + 1$, which is $2r + 1$, if we let $r = 2k^2 + 2k$. And so n^2 also is odd, as desired.   □

2.7   Prove that the product of any four consecutive positive integers is a multiple of 24.

Theorem. *The product of any four consecutive integers is a multiple of 24.*

Proof.   Suppose that we multiply together four consecutive integers $n = abcd$. Since the integers follow the pattern even-odd-even-odd, we know that two of the four numbers must be even. Indeed, since every other even number is a multiple of four, we know that one of those even numbers must be a multiple of four. So n will have to be a multiple of 8. Similarly, at least one of the four numbers will be a multiple of three, and so n also will be a multiple of 3. So n is a multiple of the least common multiple of 3 and 8, which is 24.   □

Note that 24 is the best possible for this result, in light of the fact that $1 \cdot 2 \cdot 3 \cdot 4 = 24$.

3.2 Show that a positive integer $p > 1$ is prime if and only if, whenever p divides a product ab of integers, then either p divides a or p divides b.

Theorem. *A positive integer $p > 1$ is prime if and only if, whenever p divides a product ab of integers, then either p divides a or p divides b.*

Proof. This is an if-and-only-if statement, and so we prove each direction separately.

($\rightarrow$) This direction is precisely the content of lemma 13.

($\rightarrow$) We prove the contrapositive. Assume that p is a positive integer, larger than 1, and is not prime. Thus, p has a positive integer divisor a other than 1 or p, which means $p = ab$ for some integers with $1 < a, b < p$. Thus, p divides ab, since it is equal to ab, but p does not divide a and does not divide b, since a larger positive integer cannot divide a smaller one. □

4.12 Prove that if you divide the plane into regions using finitely many straight lines, then the regions can be colored with two colors in such a way that adjacent regions have opposite colors.

Theorem. *If we divide the plane into regions using finitely many straight lines, then the regions can be colored with two colors in such a way that adjacent regions have opposite colors.*

Proof. We prove the theorem by induction on the number of lines. We might begin with the case of zero lines, in which case we may color the entire plane red. Or, if you like to interpret the problem as having at least one line, then we may anchor the induction with one line, which divides the plane into two half-plane regions, and we may color one of them red and the other blue.

Suppose that we have placed n lines in the plane and colored the regions red and blue in such a way that fulfills the requirements. Place a new line ℓ in the plane. The line ℓ divides the plane into two half planes. Let us simply swap the colors red for blue on one side of ℓ, but keep the original colors on the other side. In this way, the regions immediately cut by ℓ will exhibit the adjacent-regions-opposite-color feature, and this will also remain true for any other line, since the feature held originally, and it is preserved when you swap colors. Note that it is no problem if the new line should happen to pass through an intersection of previously occurring lines. □

An alternative proof proceeds like this: Put all the lines in the plane from the start, and then color each region red or blue, depending on whether one must respectively cross an even or odd number of lines altogether to get to that region from some fixed point not on any of the lines. This parity count is well defined, independently of the particular path of getting to that region. And adjacent regions will differ by exactly one in the parity count, and hence they will get opposite colors. □

5.3 Show that if there are infinitely many people, then it could be possible for every person to be more pointed at than pointing. Indeed, can you arrange infinitely many people, such that each person points at only one person but is pointed at by infinitely many people? How does this situation interact with the money-making third proof of theorem 34?

Theorem. *There is a pointing relation on the natural numbers such that every number points at only one other number but is pointed at by infinitely many numbers.*

Proof #1. We begin by directing all the even numbers to point at 0. Next, amongst the numbers remaining, which are the odd numbers, we direct every other one of them to point at 1. And amongst the numbers remaining after this, we direct that every other one should point at 2. And so on. In this way, it is clear that every number points at just one number, but meanwhile every number will eventually have its turn for adulation, becoming at that time pointed at by infinitely many numbers. □

Proof #2. Let us provide a more explicit pointing relation. Every positive integer has the form $n = 2^r(2k + 1)$, since we may factor the largest power of 2 out of the prime factorization of n, and what remains is odd. This form of representing n is unique, by the uniqueness of the prime factorization.

Let us direct number n to point at the exponent number r. (And let us direct 0 to point at itself.) Thus, each number points at exactly one number, but meanwhile, every number r is pointed at by all the numbers of the form $2^r(2k + 1)$, of which there are infinitely many. □

How does this theorem relate to the money-making proof of theorem 34? Well, if we had infinitely many people, each with exactly one dollar, and we directed them to pay their dollar according to the pointing scheme of the theorem, then after the payments, every person would have infinitely many dollars! This may seem paradoxical at first, since each person seems to have made a lot of money, but where did the extra cash come from? We just moved the same money around within the group. The paradox may be dispelled when one realizes that the total number of dollars held by the group has not changed—it is still the same countable infinity. Thus, although from each individual person's perspective it may seem as though the group has made money, in fact the number of dollar bills has remained unchanged.

6.4 Using the figure appearing in the chapter, prove that if you delete two opposite-color squares from an ordinary 8×8 chessboard, you can tile it with 2×1 dominoes, no matter which two opposite-color squares are omitted. Does the argument generalize to any size square grid? What about rectangular grids?

Theorem. *If one deletes any two opposite-color squares from an 8×8 chessboard, the remaining board can be tiled with 2×1 dominoes.*

Proof. Consider the 8×8 chessboard, with black lines marked as indicated in the figure.

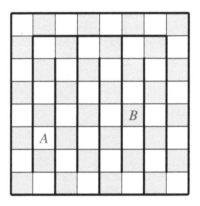

These lines depict a channel path winding over the board, visiting every square exactly once. If one were to delete two opposite-color squares, such as at *A* and *B*, or any two opposite-color squares, then this channel would be divided into two channel segments. Imagine running through the channel from *A* to *B*, and then continuing further through the channel after *B* and finding oneself at *A* again. Each of these two segments starts with one color and ends with the other, precisely because *A* and *B* have opposite colors. Thus, we may tile each of the two channel segments with 2×1 dominoes, going first from *A* to *B*, and then skipping over *B* and going from *B* to *A*. So there is a tiling of the deleted board. And it does not matter which two squares are deleted, provided that they have opposite colors.   □

We can generalize this to arbitrary rectangular boards, provided that both sides have length at least 2 and at least one side has even length.

Theorem. *On any chessboard of size $n \times m$, where these are positive integers, both at least 2, and at least one of them is even, then after deleting any two opposite-color squares, one may tile the remaining board with 2×1 dominoes. If both n and m are odd, this is never possible.*

Proof. The channel proof idea generalizes readily to larger boards. Place the even side horizontally, and define the channel as in the figure above by starting at the lower left, following a channel going up the left side, right across the top, and down on the right side to the lower right corner, and then using the up-down zigzag pattern as in the 8×8 pattern above. (Note, if $m = 2$, then there is no actual zigzagging, and the channel simply returns along the bottom edge.) Since the horizontal side length is even, and each zigzag uses two columns, this zigzag pattern will return the channel to the lower left corner as desired. And then the proof described above shows that if one deletes any two opposite-color squares, the remaining board can be tiled simply by staying within the channel.

Meanwhile, if both side lengths n and m are odd, then the board has an odd number of squares altogether, and so after deleting two squares, there are still an odd number of squares, and so it will be impossible to tile by 2×1 squares. $\qquad \square$

Note also that an $n \times 1$ rectangle also does not have the desired property, if $n > 2$, since we can delete squares one away from each edge, and then what remains cannot be tiled with 2×1 dominoes.

7.9 What is the winning strategy for misère Nim? Prove your answer.

Recall that in Nim the winning player is the one who takes the last coin, whereas in misère Nim that is the losing move.

Theorem. *The winning strategy in misère Nim is the same as for Nim, except when such a move would lead to a position with all piles having only one coin, in which case the winning misère player should instead leave an odd number of piles.*

Proof. The strategy I am proposing for misère Nim is to leave balanced positions, except when all the piles have only one coin, in which case the misère player should leave an odd number of piles. This strategy is possible to implement, since any unbalanced position can be balanced, but if the balancing move would be to act on the very last multiple-coin pile, then the player can either leave one coin or none, so as to leave an odd number of single-coin piles.

Thus, the misère player can ensure always to give either a balanced position with at least one nontrivial pile (and hence two), or a position consisting of an odd number of single-coin piles. In particular, this implies that our misère player will never take the last coin, since the empty position does not have that form, and so this is a winning strategy. $\qquad \square$

How delightful and surprising that the misère Nim strategy is so similar to the ordinary-play Nim strategy! Perhaps one might have expected it to be totally opposite in some way—many people expect at first that the misère player should be playing unbalancing moves rather than balancing moves—but the theorem shows that no, the winning moves and positions in Nim and misère Nim are exactly the same until one gets down to the case of all trivial piles.

8.9 Prove that every rectangle formed by vertices in the integer lattice (not necessarily oriented with the axes) has an area that is an integer.

In fact, we shall prove a stronger result.

Theorem. *Every parallelogram formed by vertices in the integer lattice has integer area.*

Proof. Consider any parallelogram in the integer lattice.

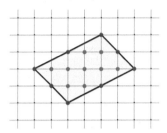

We may assume it is nondegenerate, for otherwise the area is zero, which is an integer. By Pick's theorem, the area is $i + b/2 - 1$, where i is the number of interior vertices and b is the number of lattice points on the boundary. The number i is an integer. And for a nondegenerate parallelogram, I claim, the number b is an even integer, because there are four points at the corners, and the other boundary points come in pairs, since any point on one side has a counterpart on the opposite side. So $b/2$ also is an integer, and so altogether the area is an integer.   □

> **9.8** Prove that in any rectangular lattice, using a rectangle whose side lengths are commensurable, the only regular polygon to be found using lattice points is a square.

Theorem. *The only regular polygon to be found at lattice points in a rectangular lattice, based on a unit rectangle having commensurable side lengths, is a square.*

Proof. Consider a rectangular lattice based on a unit rectangle having commensurable side lengths. Commensurability means that the sides have a common unit measure; each of them is an integer multiple of some common unit length, perhaps very small. Let us imagine constructing a square lattice using that common unit length. Because of commensurability, our rectangular lattice points will arise as points in that fine square lattice. In other words, the rectangular lattice is refined by the square lattice based on the common unit. And so any regular polygon to be found in the rectangular lattice can also be found in the square lattice. By theorem 70, the only such figures are squares.   □

> **10.2** Prove the claim in the proof of lemma 77.3 that every rectangle is dissection congruent to a rectangle with no side more than twice as long as the other.

Theorem. *Every rectangle is dissection congruent to a rectangle with no side more than twice as long as the other.*

Proof. This theorem follows, of course, from the dissection congruence theorem (theorem 77), but it would be circular for us to appeal to that theorem here, since the point was that we used this easier theorem when proving the main dissection congruence theorem. So what is called for is a separate elementary proof of this result about rectangles.

The main idea of the proof is that a long, thin rectangle can be repeatedly cut in half and restacked to make a less extreme rectangle, and by repeating this, we shall achieve the desired rectangle. Specifically, suppose that we are given a rectangle of size $a \times b$, where $b \leq a$. If $a \leq 2b$, then we are done, and so let us assume that $2b < a$, which means we have a long, thin rectangle. By slicing vertically in half, we may stack the two pieces and make a rectangle half as long and twice as high, of size $\frac{a}{2} \times 2b$. We keep doing this until for the first time the height of the rectangle is at least half the width. If we imagine having just achieved this (as pictured above), then we have $\frac{a}{4} \leq 2b$, and consequently both $2b < a$ and $\frac{a}{2} \leq 4b$, which shows that this $\frac{a}{2} \times 2b$ rectangle has both sides no more than twice the other. $\qquad\square$

11.5 Is the squaring function x^2 well defined with respect to congruence modulo 5? How about exponentiation 2^x?

Theorem. *The operation of squaring $x \mapsto x^2$ on the integers is well defined with respect to congruence modulo 5.*

Proof. Suppose that $x \equiv y \bmod 5$, which means that $y - x$ is a multiple of 5. So $y = x + 5k$ for some integer k. Therefore, $y^2 = (x + 5k)^2 = x^2 + 10xk + 25k^2$, which can be expressed as $x^2 + 5(2xk + 5k^2)$. Thus, x^2 and y^2 differ by a multiple of 5, and so $x^2 \equiv y^2 \bmod 5$, as desired. In other words, squaring a number is well-defined modulo 5. $\qquad\square$

Observation. *The exponentiation operation on the integers $x \mapsto 2^x$ is not well defined with respect to congruence modulo 5.*

Proof. Notice that 1 is equivalent to 6 modulo 5 but that 2^1 is not equivalent to 2^6, since $2^1 = 2$ and $2^6 = 64 \equiv 4 \bmod 5$. So the operation of exponentiation does not always map equivalent numbers to equivalent numbers, and so it is not well defined with respect to this equivalence relation. $\qquad\square$

12.2 Prove that in any finite graph, the total degree, meaning the sum of the degrees of all the vertices, is always even. Conclude as a corollary that no finite graph can have an odd number of odd-degree vertices.

Theorem. *In any finite graph, the total degree is twice the number of edges. In particular, it is even.*

Proof. Consider any finite graph, with a set of vertices and a set of edges between them. The total degree is the sum of the degrees of each vertex. Each edge connects one vertex to another (or to the same vertex) and therefore makes a contribution of exactly 2 to the total degree. So the total degree is simply twice the number of edges. And so it is even.   □

Corollary. *No finite graph can have an odd number of odd-degree vertices.*

Proof. If a graph had an odd number of odd-degree vertices, then the total degree would be odd, since the sum of any number of even numbers and an odd number of odd numbers is odd. But the theorem shows that the total degree in a finite graph is always even, and so this cannot happen.   □

13.6 Prove or refute: A set A is countable if and only if there is a surjective function $f : \mathbb{N} \to A$. If this is false, state and prove a closely related true theorem.

Claim. *The statement is false. There is a countable set A, with no surjection from $\mathbb{N}$ to A.*

Proof. This is a very picky point, but we can make a counterexample with the empty set $A = \emptyset$. The empty set is countable because it is bijective with a set of natural numbers (indeed, it already *is* a set of natural numbers, the empty set of numbers). But there is no function $f : \mathbb{N} \to \emptyset$ at all, since there is no possible value of $f(0)$.   □

Meanwhile, we can prove the positive result by adding the relevant hypothesis.

Theorem. *A set A is countable and nonempty if and only if there is a surjective function $f : \mathbb{N} \to A$.*

Proof. Suppose that A is nonempty and countable. Since it is countable, there is a bijection between A and a set of natural numbers $g : A \to B \subseteq \mathbb{N}$. Pick a particular element $a \in A$, which is possible since A is nonempty. Let $f(n) = g^{-1}(n)$ if $n \in B$, and otherwise $f(n) = a$. Since g is bijective, every element of A has the form $g^{-1}(n)$ for some $n \in B$, and so f is a surjection from $\mathbb{N}$ to A.

Conversely, suppose that there is a surjection $f : \mathbb{N} \to A$. It follows that A is nonempty. For each $a \in A$, let $g(a)$ be the smallest number n with $f(n) = a$. Such an n exists since f is surjective. Different elements a will necessarily get different numbers n, and so g is injective. Thus, g is a bijection of A with its range, which is a set of natural numbers. So A is countable.   □

14.8 Suppose that ≤ is a partial order. Define that elements p and q are *comparable* if either $p \leq q$ or $q \leq p$; otherwise, they are *incomparable*. Prove or refute: Comparability is an equivalence relation. Does your answer change if you know that the order is linear?

Theorem. *The comparability relation in a partial order is not necessarily an equivalence relation. While it is always reflexive and symmetric, there are partial orders whose comparability relation is not transitive.*

Proof. In any partial order, every element is comparable to itself, and if a and b are comparable, then so are b and a. So comparability is both reflexive and symmetric.

But it need not necessarily be transitive. To see this, consider the partial order specified by the following Hasse diagram:

Notice that a is comparable to c, and c is comparable to b, but a and b are not comparable. This is a violation of transitivity for the comparability relation in this order. So comparability in this order is not an equivalence relation. □

But notice that all elements in a linear order are comparable (indeed, that is the definition of linearity), and so the comparability relation in a linear order is the complete (always true) relation, which is an equivalence relation.

15.10 Prove that every triangle in the plane has a line that bisects both the area and the perimeter simultaneously. Is it necessarily unique?

Lemma. *For any triangle T in the plane and any angle θ, there is a line ℓ making angle θ with respect to the x-axis and bisecting the area of T, and furthermore, this line is unique.*

Proof. Consider a fixed line ℓ_0 at the desired angle but passing completely below the triangle. Let ℓ_t be the line obtained by vertically displacing ℓ_0 by distance t. There is some large enough distance k so that ℓ_k lies totally above the triangle. Let $f(t)$ be the fraction of the area of T below ℓ_t. So $f(0) = 0$ and $f(k) = 1$. Notice that the function f is continuous, since by insisting that displacements are very tiny, we can thereby ensure that the change in the area is as small as we like, and consequently, the change in f can also be guaranteed to be as small as we like. By the intermediate-value theorem, therefore, there is some value of t for which $f(t) = 1/2$. In other words, line ℓ_t exactly bisects the area of the triangle, as desired. This area-bisecting line is unique, since any larger value of t will have too much area below, and any smaller value of t will have too much area above.   □

Theorem. *Every triangle in the plane has a line that bisects both the area and the perimeter simultaneously.*

Proof. For each angle θ, let $\ell(\theta)$ be the unique line that makes angle θ with the x-axis and bisects the area of the triangle. For each angle θ, we view the line $\ell(\theta)$ as having a right side and a left side, as determined by one's right and left hands, if one should follow the angle θ out of the origin. For example, when the angle is 0, then we use the portion below $\ell(0)$ as the right side, and for $0 < \theta < 90°$, we continue to use the portion below $\ell(\theta)$ as the right side. When $90° \leq \theta < 180$, this is better described as the half plane to the right or above $\ell(\theta)$. At $\theta = 180°$, we are naturally using the upper half plane as the "right" side.

Let $p(\theta)$ be the proportion of the perimeter of the triangle that is on the right side of the line $\ell(p)$. Notice that this is a continuous function, since if some small change in the proportion of the perimeter is desired, one can ensure this by making only sufficiently small changes in the angle.

Finally, the key thing to notice is that, since $\ell(\theta)$ and $\ell(\theta + 180°)$ are the same line but with their right and left sides swapped, it follows that $p(\theta + 180°) = 1 - p(\theta)$. In particular, if $p(\theta) < 1/2$, then $p(\theta + 180°) > 1/2$, and so in any case, by the intermediate-value theorem, there must be some angle α with $p(\alpha) = 1/2$. Therefore, the line $\ell(\alpha)$ bisects both the area and the perimeter of the triangle.   □

Finally, we note that the simultaneous-bisecting line need not be unique. For example, an equilateral triangle has three such lines, the angle bisectors of any of the corners.

Bibliography

Bouton, Charles L. 1901. Nim, a game with a complete mathematical theory. *Annals of Mathematics* 3 (1/4): 35–39. doi:10.2307/1967631. http://www.jstor.org/stable/1967631.

Cantor, Georg. 1874. Über eine Eigenschaft des Inbegriffs aller reelen algebraischen Zahlen. *Journal für die Reine und Angewandte Mathematik (Crelle's Journal)* 1874 (77): 258–262. doi:10.1515/crll.1874.77.258.

Cantor, Georg. 1891. Ueber eine elementare Frage der Mannigfaltigkeitslehre. *Jahresbericht der Deutschen Mathematiker-Vereinigung* 1: 75–78. http://gdz.sub.uni-goettingen.de/pdfcache/PPN37721857X_0001/PPN37721857X_0001___LOG_0029.pdf.

Chao, Yuen Ren. 1919. A note on "continuous mathematical induction." *Bulletin of the American Mathematical Society* 26 (1): 17–18. https://projecteuclid.org:443/euclid.bams/1183425067.

Clark, Pete. 2019. The instructor's guide to real induction. *Mathematics Magazine* 92: 136–150. doi:10.1080/0025570X.2019.1549902.

Climenhaga, Vaughn. 2017. Proofs without words. MathOverflow answer. https://mathoverflow.net/q/25305.

Dickman, Benjamin. 2013. Elementary+Short+Useful. MathOverflow answer. http://mathoverflow.net/q/152300.

Dickman, Benjamin. 2017. Enriching divisibility: Multiple proofs and generalizations. *Mathematics Teacher* 110: 416–423. doi:10.5951/mathteacher.110.6.0416.

Golomb, S. W. 1954. Checker boards and polyominoes. *American Mathematical Monthly* 61: 675–682. doi:10.2307/2307321. https://ezproxy-prd.bodleian.ox.ac.uk:2095/10.2307/2307321.

Larson, Loren. 1985. A discrete look at $1 + 2 + \cdots + n$. *College Mathematics Journal* 16 (5): 369–382.

O'Connor, Russell. 2010. Proofs without words. MathOverflow answer. http://mathoverflow.net/q/17347.

Pick, Georg. 1899. Geometrisches zur Zahlenlehre. *Sitzungberischte des Deutschen Naturwissenschaftlich-Medizinischen Vereines für Böhmen "Lotos" in Prag*, 47: 315–323.

Pólya, G. 1973. *How to Solve It: A New Aspect of Mathematical Method*, 2nd edn. Princeton University Press.

Pritchard, Dave. 2010. Proofs without words. MathOverflow answer. http://mathoverflow.net/q/24828.

Shreevatsa. 2010. Proofs without words. MathOverflow answer. http://mathoverflow.net/q/17328.

Silagadze, Zurab. 2014. Proofs without words. MathOverflow answer. http://mathoverflow.net/q/163807.

Suárez-Álvarez, Mariano. 2009. Proofs without words. MathOverflow question. http://mathoverflow.net/q/8846.

User3203476. 2016. Why is this "proof" by induction not valid? Mathematics Stack Exchange question. http://math.stackexchange.com/q/2066805.

Wayte, William. 1878. Annotations on game 119. *Chess Player's Chronicle* 2 (13): 30–31.

Wikipedia. 2020. Seven Bridges of Königsberg. https://en.wikipedia.org/wiki/Seven_Bridges_of_K%C3%B6nigsberg.

Index of Mathematical Habits

The reader has likely noticed that every chapter in this book ends with a collection of mathematical habits of mind, highlighted in boxes. These are bits of mathematical wisdom or suggestions that I have gathered, intending them as general advice for young mathematicians. For your convenience, I have assembled an index of all the "habits" here.

Notation Index

Each index entry refers to the first or most prominent use in this book of a particular mathematical notation.

Subject Index